SpringerBriefs in Computer Science

Series editors

Stan Zdonik
Shashi Shekhar
Jonathan Katz
Xindong Wu
Lakhmi C. Jain
David Padua
Xuemin (Sherman) Shen
Borko Furht
V.S. Subrahmanian
Martial Hebert
Katsushi Ikeuchi
Bruno Siciliano
Sushil Jajodia
Newton Lee

More information about this series at http://www.springer.com/series/10028

Umit Isikdag

Enhanced Building Information Models

Using IoT Services and Integration Patterns

 Springer

Umit Isikdag
Mimar Sinan Fine Arts University
Istanbul
Turkey

ISSN 2191-5768   ISSN 2191-5776   (electronic)
SpringerBriefs in Computer Science
ISBN 978-3-319-21824-3   ISBN 978-3-319-21825-0   (eBook)
DOI 10.1007/978-3-319-21825-0

Library of Congress Control Number: 2015945143

Springer Cham Heidelberg New York Dordrecht London

Printed on acid-free paper

Springer International Publishing AG Switzerland is part of Springer Science+Business Media (www.springer.com)

Preface

Have you ever thought of the Internet as a 'Thing'? A physical object that you can hold, measure the dimensions, visualize and so on. You may suggest that the Internet is a combination of physically and non-physically existent 'Things' such as communication rules, messages, information sent from A to B, which is also true. But how many of you think of cables and satellites, when you send an e-mail or start a video conference? Actually we use a 'Thing' to do that, a global 'Thing' that has physical and non-physical components. However, regardless of the technology behind it, we concentrate on if it gets the job done, and mostly it does. Thus the main focus is not on the 'Thing' itself, but on information, and it concerns the success of the sharing and exchange of information. Do you think this vision is enough? Maybe not, we also want to receive information as soon as something happens. We want real-time information. Actually we do not care too much about physical 'Things', but we do care about the states of 'Things'. We are curious. We would like to learn what is happening all around us. As soon as possible!

The key technologies we elaborate on in this book are the Internet of Things (IoT), Web services and building information modelling. The first technology, IoT, aims to answer the questions discussed until now. IoT does not care about the existence of 'Things'. 'Things' can be real, 'Things' can be virtual; what IoT really focuses on is the state of 'Things'. The approach concentrates on making every physical and virtual 'Thing' a publisher of information, like the nerve cells in the brain. The IoT approach enables 'Things' to publish information when a state change occurs. For instance, in a home that implements the IoT approach, a door will publish information such as 'I am locked now!', a light bulb will indicate 'I am in a morning blue color at the moment'. Are these the only cases of all the hype about the IoT? There are more. The 'Things' will also become capable of taking actions based on messages coming from other 'Things' or humans. For example, you can use a 'Thing' (a cell phone) to control your home lighting while you are far away, or you can turn your TV set off from another country when you think it is time for the children to sleep. These ideas were the science fiction of yesterday, but are the science of today; a reality that has been a part of our lives for just a few years, but will be in our lives for many more. It is inevitable that the technologies termed within the context of IoT will be a part of our lives. This is so with the other two technologies that this book focuses on. Web services for one. The Internet came into our everyday life around 20 years back. At that time it was viewed as a new way of speaking with friends, new way of sending mails, a new way of marketing and selling goods and a new way of expressing oneself to the world. Over the past 20 years, although we are still confronted with issues of digital divide, things have significantly changed. For instance, mobile devices are now of no use if they cannot connect to the Internet. It is the same with tablet computers. The question is how do we interpret a situation where the role of a single technology, such as the Internet or a 'Thing' becomes useless if it does not benefit from a certain technology. Let us take the analogy of electricity/water and dishwashers. 'Things' need to benefit from utilities in order to work; however, once a technology comes to the level that a 'Thing' cannot work without that technology, the latter is no more a technology but a utility. The situation is the same today for the Internet. The Internet will become a key utility in the future. From this perspective Web services can be thought of as interaction endpoints of this utility. Today, there are architectural advancements on the implementation of these endpoints (such as Representational State Transfer). In fact it should be noted that these endpoints are not entry/exit points (such as plugs for electricity), but they enable us to interact with (hardware/software) components that make use of this utility. Thus, Web services are endpoints for interaction. It is our choice actually to use these endpoints (or not) for interaction, as there are also other choices that we can use such as sending messages from one component to another, or from a human to the components. Message brokers are middleware tools that help us to distribute these messages. Finally, building information models is another hype that has been a buzzword in the construction industry for the past 15 years. These models have emerged as a result of a thrust by software companies to tackle problems of inefficient information exchange between different softwares and to enable true interoperability. An industry standard schema (namely Industry Foundation Classes) was developed to facilitate information exchange between construction industry applications. Later, the industry noted that models produced within a common schema could be utilized to enable shared use of information with the help of shared databases. Thus, BIM became the data sharing technology, where the most up to date and accurate models of a building are stored in shared central databases. This opened new doors. Industry started to focus on making pre-construction simulations using these models, accompanied by multiple stake-holders, which is now termed the nD modelling approach. Later, the information residing in the models was maintained following the construction phase, and the models started to act as the virtual ID cards of the buildings. In parallel, devel-opments in city modelling led to information requirements from these models, which have now become the information providers of the digital city. The city is a living entity and city-level applications require information from 'Things' (i.e. real and virtual) and from 'Models' in real time. Thus, today emerges the requirement for real-time information regarding buildings, indoors and all other city elements in order to efficiently monitor and manage a city. In essence, the construction industry applications (such as smart buildings) and city monitoring/city management applications require the fusion of information acquired from multiple resources, 'Things', models, virtual objects and real objects.

This book focuses on providing approaches and software architectures for (i) facilitating interaction with building information models through Web services and (ii) enabling and facilitating the fusion of building information residing in 'Models' and information acquired from the 'Things'. The proposed architectures are presented in the form of design patterns. The patterns utilize IoT technologies, Web services and BIMs. Once this information fusion is accomplished, many fields ranging from emergency response, urban surveillance, urban monitoring to smart buildings will benefit. The book will be beneficial for researchers and developers in the fields of building information modelling, IoT and systems integration. The book consists of eight chapters. Chapters 1 and 2 focus on building information modelling. Chapter 3 provides foundational service-oriented architecture patterns (SOA) for complex information models (such as BIM); in fact, the implementation of these patterns can also be accomplished using other information models. Chapters 4 and 5 elaborate on the hardware and software sides of IoT. Chapter 6 provides advanced SOA patterns for BIMs. Chapter 7 elaborates on patterns for IoT and patterns for BIM and IoT information fusion.

This book has taken a whole year to complete, although the actual schedule was 7 months. I would like to thank my wife Zeynep Isikdag for her patience and support during the process and for many weekends during which we stayed at home, and to my parents, Dr. Ugur Isikdag and Zuhal Isikdag, for always supporting me in my academic life. I thank all academics, who supported me and from whom I have learned a lot, including my Ph.D. supervisors Jason Underwood, Ghassan Aouad, Nigel Trodd, and Sisi Zlatanova, Volker Coors, who contributed a lot during my Ph.D. Thanks also to my M.Sc. supervisors Keith Jones and Robert Falconer. Special thanks to all members of the 3D GeoInfo Community, who have always supported me, and especially to the initiator of the 3D GeoInfo Conference, Alias Abdul Rahman. Also, special thanks to the editorial board members of the International Journal of 3D Information Modeling (IJ3DIM), I learned a lot from their knowledge and experience, and also to Jack Goulding for his academic support in the construction management field. I would also like to thank Ralf Gerstner from Springer for his excellent support and patience, without whom this book could not be possible.

May 2015   Umit Isikdag

Contents

1 Building Information Models: An Introduction 1
  1.1   Introduction 1
  1.2   Defining Building Information Modelling 3
  1.3   Industry Foundation Classes Model 5
  1.4   Storage and Exchange of Building Information Models 6
  1.5   Views of Building Information Models 7
  1.6   The Role of BIM in the Enterprise 8
  References 11

2 The Future of Building Information Modelling: BIM 2.0 13
  2.1   Introduction 13
  2.2   Research Dimensions of Building Information Modelling 14
      2.2.1   Information Model-Related Aspects 15
      2.2.2   Organizational Aspects 16
      2.2.3   Domain-Specific Aspects 17
      2.2.4   Project Management Aspects 17
      2.2.5   Integration and Interoperability Aspects 17
  2.3   BIM-M: Utilization of BIM in Construction Management 18
  2.4   Technologies for BIM 2.0 19
  2.5   BIM-Based Management of Construction Processes 20
  References 23

3 Foundational SOA Patterns for Complex Information Models 25
  3.1   Introduction 25
  3.2   Design Principles of Service Orientation 26
  3.3   Complex Information Models 28
  3.4   Service-Oriented Patterns 30
      3.4.1   Data Definition Language Provider 30
      3.4.2   Model View Selector 31
      3.4.3   Model View Entity Extractor 32
      3.4.4   Sub-view Generator 33

        3.4.5   View Observer...............................   34
        3.4.6   View Updater   35
        3.4.7   Extended Model Observer......................   36
        3.4.8   Extended Model View Observer.................   37
        3.4.9   Extended Model View Updater   38
        3.4.10   Model Controller   39
    References   41

4 **Internet of Things: Single-Board Computers**   43
    4.1   Introduction.....................................   43
    4.2   Arduino Development Boards   44
    4.3   BeagleBoard   46
    4.4   CubieBoard   47
    4.5   Raspberry Pi   48
    4.6   Orange Pi   49
    4.7   UDOO Board......................................   49
    4.8   Netduino Board   50
    4.9   Intel Galileo and Edison   51
    4.10   Radxa Rock......................................   52
    References   53

5 **Internet of Things: Software Platforms**   55
    5.1   Introduction.....................................   55
    5.2   Operating Systems   56
        5.2.1   Mobile Operating Systems.....................   56
        5.2.2   OpenWRT....................................   57
        5.2.3   Windows Embedded...........................   57
        5.2.4   Raspbian....................................   57
        5.2.5   Contiki OS   57
        5.2.6   RIOT OS   58
        5.2.7   Tiny OS   58
        5.2.8   Free RTOS   58
    5.3   Hardware and Software Bundles......................   59
        5.3.1   Spark.IO   59
        5.3.2   Open Mote   59
    5.4   Messaging Standards and Protocols....................   60
        5.4.1   RPL Protocol................................   60
        5.4.2   6LoWPAN Protocol   60
        5.4.3   CoAP Protocol...............................   61
        5.4.4   MQTT Protocol   61
        5.4.5   XMPP Protocol   61
    5.5   Middleware and Frameworks.........................   62
        5.5.1   AllSeen Alliance and AllJoyn   62
        5.5.2   Eclipse IOT Frameworks and Services   62

5.5.3   IoTSyS Middleware   63
5.5.4   IoTivity Framework   63
5.5.5   OpenIoT Project...........................   63
5.5.6   Macchina.IO   64
5.6   Integration Portals   64
5.6.1   Xively...................................   64
5.6.2   Paraimpu.................................   67
5.6.3   Dweet.IO.................................   68
5.6.4   Freeboard.IO   69
References   69

6   **Advanced SOA Patterns for Building Information Models**   71
6.1   Introduction...................................   71
6.2   REST in a Nutshell   72
6.3   Generalized Design Pattern for BIM-Based SOA   76
6.4   REST Query Filter Pattern..........................   79
6.5   REST Façade Pattern   81
6.6   RESTful Real-Time View Generator Pattern   82
6.7   RESTful Memento Pattern..........................   84
6.8   RESTful Model Multi-view Controller Pattern   86
6.9   RESTful Call-Back Responder Pattern   88
6.10   RESTful Authenticator Pattern......................   90
6.11   RESTful Data Management Pattern....................   91
6.12   RESTful View Synchronizer Pattern   93
6.13   RESTful Event Manager Pattern......................   95
References   97

7   **Sensor Service Architectures for BIM Environments**   99
7.1   Introduction...................................   99
7.2   Sensor and BIM Integration Patterns..................   101
7.3   Foundational Publish-Subscribe   101
7.4   Feed Encoder..................................   102
7.5   Message-Based Cloud Update   104
7.6   On-Demand Cloud Update..........................   106
7.7   RESTful Node Façade.............................   108
7.8   BIM and IoT Service Façade   109
7.9   BIM Updater Nodes   111
7.10   Rich Client for BIM and IoT Nodes   112
7.11   Real-Time BIM Callback...........................   113
7.12   BIM Virtual Sensors.............................   114
References   115

8 **Summary and Future Outlook** 117
    8.1   Overall Summary 117
    8.2   Future Outlook 119

Chapter 1
Building Information Models: An Introduction

Abstract A building information model (BIM) can be defined as the digital representation of a building that contains semantic information about the building elements. The keyword BIM also defines an information management process based on the collaborative use of semantically rich 3D digital building models in all stages of the project's and building's lifecycle. A BIM is defined by its object model schema. Industry Foundation Classes (IFC) is the most popular BIM standard (and schema) currently. This chapter starts by providing definitions of BIM and the general characteristics of IFC models, elaborates on sharing/exchanging of BIMs and model views, and concludes by discussing the role of BIMs in enterprises.

1.1 Introduction

Today, building information modelling is an active research area that addresses problems related to sharing of information, interoperability and efficient collaboration throughout the lifecycle of a building (i.e. from feasibility and conceptual design through to demolition and re-cycling stages). Building information modelling has its roots in two distinct but closely related fields. The first field is computer-aided design (CAD) which focuses on designing using computers; the second is 'Representation of Building Product Information' which focuses on providing information related to the components of a building in an organized way. CAD applications were among the first computing applications used in the construction industry. In the 1970s, university-led CAD research projects aimed to create design drawings and building performance measures. Energy saving became an important issue after 1973 due to the rise in oil prices. Heat loss calculations are attached to CAD systems, as environmental effects are linked with the shape of the

© The Author(s) 2015
U. Isikdag, *Enhanced Building Information Models*,
SpringerBriefs in Computer Science, DOI 10.1007/978-3-319-21825-0_1

building. In the mid-1990s, AutoDesk's DXF format became a de facto standard for the exchange of 2D geometric information. The main drawback of the CAD outputs was that the drawings in CAD documents only consisted of sets of polylines and polygons which did not contain semantic and ontological (i.e. product) information about the building and its components. Parallel to the developments in the CAD domain, the efforts towards the representation of building product information generated results such as definition of classification systems for materials. Classification systems such as OMNICLASS and UNIFORMAT later served as the foundation for the idea of building product models (where semantic information is stored together with geometric information). According to Tolman (1999), early product models included general AEC reference model (GARM) (Gielingh 1988), integration core model (ICM) and the integration reference model architecture (IRMA). The idea behind the definition of building product models was to facilitate the representation of building-related product information at the most appropriate time and to the right project team member. Building product models had the following characteristics:

- They provided detailed geometric and semantic information about all building elements in a tightly coupled form.
- They focused on addressing the problem of poor interoperability between software applications in the construction industry.
- Most of them were defined based on ISO 10303 data definition guidelines.

ISO 10303, which is also known as STEP (STandard for Exchange of Product model data), emerged in the early 1990s as a formal standard to exchange product data in all production industries. The emergence of STEP was a result of the issues associated with the shortcomings of CAD data translation. The distinction between data sharing and exchange is clearly identified during STEP development efforts; in addition, the STEP standard identified four implementation levels for data storage and exchange. These will be explained further in this chapter. The early efforts for building product modelling continued with the development of the building construction core model (BCCM), which was later approved as Part 106 of the STEP standard (ISO 10303). Another early effort in the area included the COMBINE project, which was explained by Sun and Lockley (1997) and Eastman (1999). Other important efforts in the area included computer integrated manufacturing of constructional steelwork—CIMSteel and CIMSteel integration standards (CIS/2)—explained in Eastman (1999), NIST CIS/2 Website (2005), and Eastman et al. (2005), engineering data model (EDM) as explained by Eastman et al. (1991), semantic modelling extension (SME) explained in Zamanian and Pittman (1999), models developed in the integrated design environment (IDEST) project as explained in Kim et al. (1997), Kim and Liebich (1999), RATAS and STEP Part 225 as explained in Eastman (1999). These efforts later continued with the introduction of the IFC and CIS/2 building product models, which formed the basis for the paradigm that is known today as building information modelling. First, we look at the definition of building information modelling; later we elaborate on the IFC model architecture, the role of model views and the function of BIM in enterprises.

1.2   Defining Building Information Modelling

The traditional nature of the construction industry is 'document-centric'. The documents contain drawings, regulations, specifications and so on. The form of information stored in documents is rich and multidimensional. When parties in the construction process are required to exchange or share information using documents, immense barriers to communication are faced among the various stakeholders, which in turn significantly affects the efficiency and performance of the industry. Gallaher et al. (2004) indicated that US$15.8B is lost annually in the U.S. capital facilities industry due to lack of interoperability (i.e. caused by inefficient exchange of information). Building information modelling focuses on overcoming problems related to information exchange and sharing throughout the lifecycle of the building. Today, building information models aim to be the facilitators of integration, interoperability and collaboration in the construction industry. According to NBIMS (2006), a BIM is a computable representation of all the physical and functional characteristics of a building and its related project/lifecycle information, which is intended to be a repository of information for the building owner/operator to use and maintain throughout the lifecycle of a building. The US General Services Administration BIM Guide (2006) indicates that the information in a BIM catalogues the physical and functional characteristics of the design, construction and operational status of the building. BIM as a process can be defined as the information management process, which mainly focuses on enabling and facilitating the integrated way of project flow and delivery by collaborative use of semantically rich 3D digital building models in all stages of the project and building lifecycle. The BIM process is unique as it is based on digital, shared, integrated and interoperable building information models. From this perspective, the building information modelling process can be defined as a facility that enables information management throughout the lifecycle of a building, while a building information model is the (set of) semantically rich shared 3D digital building model(s) that form(s) the backbone of the building information modelling process. Based on the literature in the field, the definitive characteristics of BIM are identified as follows:

1. *Object oriented*: Most BIMs are defined in an object-oriented manner.
2. *Open/Vendor neutral*: BIMs are developed for effective information exchange and sharing; therefore, being open/non-proprietary and vendor neutral is recognized as an important characteristic.
3. *Enables interoperability*: BIMs are developed to overcome the problem of insufficient interoperability; thus, it is recognized as a natural characteristic.
4. *Data rich/Comprehensive*: BIMs are data rich and comprehensive as they cover all physical and functional characteristics of the building.
5. *Extensible*: BIMs can be extended to cover different aspects of the information domain.

6. *Three dimensional*: BIMs always represent the geometry of the building in three dimensions.

7. *Covers various phases of the project lifecycle*: The state-of-the-art BIMs cover various phases of the project lifecycle. The objects of the model can be in different states in different phases of the lifecycle in order to represent the n-dimensional information about the building.

8. *Spatially related*: Spatial relationships between building elements are maintained in BIMs in a hierarchical manner.

9. *Rich in semantics*: BIMs store a large amount of semantic (functional) information about the building elements.

10. *Supports view generation*: The model views are subsets or snapshots of the model that can be generated from the base information model. BIMs therefore support view generation.

11. *Stored, shared and exchanged*: BIMs can be stored as a file or in a database, be shared in databases or with the help of APIs (when it is a physical file) and can be exchanged in the form of physical files.

Building information modelling is applied in different areas and at different levels; that is, either the models are used as a resource to enable interoperability or building information modelling is realized as a process of managing a project through a single shared information backbone. Recent research in the area demonstrated how n-dimensional simulation applications can be facilitated using BIMs (Rebolj et al. 2010; Spearpoint 2010). Countries such as Singapore use BIMs to validate that building models are compliant with the code and regulations. Research has also demonstrated that building information modelling can facilitate the design of energy-efficient buildings towards addressing sustainability and reduction in CO_2 emission issues (Solis and Mutis 2010; Hua 2010).

BIMs play the well-recognized role of facilitating the design phase of a project lifecycle. Depending on the environment in which they are used, BIMs can have different functions such as being a space linker that links macro and micro urban spaces (i.e. city and building level), an interoperability enabler that facilitates information sharing between various stakeholders and the software applications they use, a data store that stores the building information throughout the lifecycle of a building, a procurement facilitator that facilitates several procurement-related tasks in the building lifecycle, a collaboration supporter through enabling the use and management of shared building information in real-time, a process simulator by facilitating the simulation of construction processes (i.e. nD), a system integrator that enables the integration of several information systems across the industry, a building information service that can serve real-time on-demand building information over the Internet, a green builder that enables advanced analysis supporting the design and construction of environment friendly/energy-efficient buildings and a life saver that facilitates emergency response operations.

1.3   Industry Foundation Classes Model

The current key efforts in the area of BIM are IFC and CIS/2, both of which are defined using STEP (ISO 10303) description methods. In this section, we focus on IFC, which is the most popular BIM standard (and schema) currently. IFC appeared as a result of the effort of IAI/BuildingSmart as a common language to improve the communication, productivity, delivery time, cost and quality throughout the design, construction and maintenance of buildings. In the IFC model, each specification (called a 'class') is used to describe a range of things that have common charac-teristics. IFC-based objects aim to allow AEC/FM professionals to share a project model while allowing each profession to define its own view of the objects con-tained within the model. In 2005, IFC became an ISO Publicly Available Specification (as ISO 16739). Most AEC industry software today is capable of importing and exporting its internal models as IFCs, and some of them are also capable of acquiring information from an IFC model through the use of a shared resource such as a model server database.

The IFC model architecture provides a modular structure for the development of model components by use of 'model schemata'. There are four conceptual layers within the architecture (see Fig. 1.1) which use a strict referencing principle. Within each conceptual layer a set of model schemata is defined.

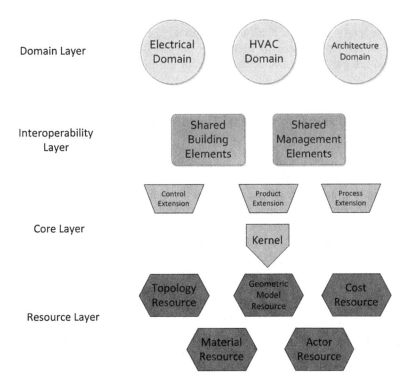

Fig. 1.1 IFC 2x4 components subset

1. The first conceptual layer (resource) provides resource classes used by classes at the higher levels.
2. The second conceptual layer (core) provides a core project model and contains the kernel and several core extensions.
3. The third conceptual layer (interoperability) provides a set of modules defining concepts or objects common across multiple application types or construction industry domains.
4. Finally, the fourth and the highest layer (domain) provides a set of modules tailored for the specific construction industry domain or application type.

The architecture operates on a 'gravity principle'. At any layer, a class may reference a class at the same or at the lower layer, but may not reference a class from a higher layer.

Resources are general-purpose or low-level concepts or objects that are independent of application or domain, but which rely on other classes in the model for their existence. The core layer provides the basic structure of the IFC object model and defines most general concepts specialized by higher layers of the IFC object model. The main goal of the design of the interoperability layer is the provision of schemata that defines the concepts (or classes) common to two or more domain models. These schemata enable interoperability between different domain models. Domain layer models provide further model details within the scope of requirements for a domain process or for a type of application. An important purpose of domain models is to provide the leaf node classes that enable information from external property sets to be attached appropriately. Figure 1.1 shows a sample of selected components/classes in each layer of the IFC 2x4 model. Readers are advised to refer to IFC 2x4 documentation for the full model reference.

1.4 Storage and Exchange of Building Information Models

The transfer of information between different systems is and continues to be an apparent need in the construction industry as the industry is highly fragmented and there are a variety of different information systems that are used in each organization. BIMs have been developed with the objective of providing exchange and sharing of information between different software applications. These may be either of the same type—homogeneous applications (i.e. CAD to CAD)—or of different types—heterogeneous applications (i.e. CAD to thermal load calculation, etc.) BIMs of today such as IFC and CIS/2 are compliant with the STEP standard, which means the model can be shared and exchanged using sharing and exchange methods that are originally defined by STEP (ISO 10303). As indicated by Kemmerer (1999) and Schenk and Wilson (1994), the characteristics that distinguish data sharing from data exchange are the centrality and ownership of that data. In the exchange model, one software system maintains the master copy of the data internally and exports a snapshot of the data for others to use. Other software systems that import the exchange file have effectively assumed the ownership of the data. In the sharing model, there is a centralized control of ownership and there is a known master copy of the data, i.e. the copy maintained by the information resource. STEP has four different implementation levels derived from PDES implementation levels. Loffredo (1998) mentioned these four levels as file exchange level, working form level (SDAI API access), database level and knowledgebase level. The levels mentioned here, except the last one, are for the exchange and sharing of BIM, but there are also further novel methods that are in use today for BIM information sharing, as BIM of today are also defined/described as non-compliant with STEP description methods (i.e. by using a non-STEP compliant XML language).

A BIM is defined by its object model. The object model of the BIM is the logical data model that defines all entities, attributes and relationships. The object model today is implemented in the form of EXPRESS or XSD schemas. The model data (i.e. instances) is created by an application (e.g. CAD, analysis, etc.) and stored in physical files or databases. As mentioned above, it is possible to share and exchange BIMs using three implementation levels of STEP, if the model is defined using STEP description methods (such as EXPRESS). If not, it will possibly be defined and made popular as a model in an XML file or in a relational or object database, and the data will either be exchanged as XML files or sharing will be realized using the XML database interfaces.

1.5   Views of Building Information Models

In order to support the several phases and stakeholders of construction lifecycle, several views of the BIM need to be generated. These views can be generated from files or databases using application, database and Web interfaces. The views can either be transient or persistent depending on the need and generated by a declaration of a model that is a subset of another model, or by a declaration of a model that is derivable from another model. The original model is called the base model and the new model is called the view. The entities of the view are populated from the base model.

The persistent model views are generated by model translation, and under the following conditions the (translated) model can be called the model view:

1. The view should not be a superset of a predefined (base) information model. The view can be a subset of the model or the model itself.
2. The view should provide a snapshot of the information model (or its subset).

If the model view is persistent, then it will be stored in a physical file or in a database; however, if it is transient the physical storage of the view is not necessary. The persistent model views can be used whenever there is a need to exchange a subset of BIM between various domains or when there is a need to exchange a snapshot of the BIM in one stage of the project. Another type of model view is the application/system-specific view. The application/system-specific view does not have to be a subset of the base information model. By contrast, this view is an information model of its own, defined according to the needs of the application/system it works with. As mentioned in the literature, an information model can be called an application/system-specific view based on the following conditions:

1. The model should interact with a base information model.
2. The model should address the specific data needs of an application/system it is developed for.
3. The model should address a similar information domain with the base information model.
4. The model should address the same information domain as the application/system it is developed for.

In common practice, the model views (transient, persistent and application specific) are generated and updated using STEP EXPRESS-X and XSL languages.

1.6   The Role of BIM in the Enterprise

In the past 20 years, several research projects envisioned the role of building information models in enterprise (software) architectures. As mentioned above, they emerged to facilitate interoperability of software, but later building information modelling gained more popularity as a process of information management through the lifecycle of a building or a facility. BIM paradigm goals such as efficient information exchange, better coordinated design and construction, construction process simulations in higher dimensions (4D time, 5D cost, …) are now being realized not only by use of BIMs alone but together with many other supporting tools and technologies. In the beginning of the 2000s, the role of BIM was perceived as being a common shared repository of information for the construction enterprise where different software in use can acquire information from and update information to the model. For example, the participants of the EU research project ROADCON (ROADCON Project Deliverable 4 and 5.2 2001) explained that a BIM can be created with an architectural design application and the structural design can be carried out using the same model. Likewise, heating, ventilation and air conditioning (HVAC) and electrical and lighting designs can be undertaken using the same model. 4D simulations can be made to evaluate several different phases of construction. The model will also contain information about materials and their properties, and the facilities management (FM) services can benefit from the model after the construction phase (see Fig. 1.2).

The envisioned role of BIMs is not only limited to being a shared information resource; as explained in the project, many services will also support the role of BIM in the overall construction enterprise. Different levels were identified in the

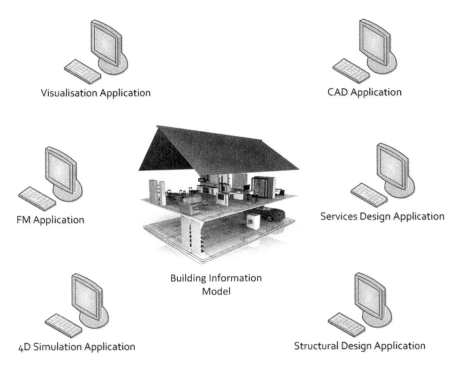

Visualisation Application

CAD Application

FM Application

Services Design Application

Building Information
Model

4D Simulation Application

Structural Design Application

Fig. 1.2 Software interactions with a building information model (image in the *middle* is courtesy of njaj at FreeDigitalPhotos.net)

ROADCON project, where stakeholders could interact with the construction information using BIMs (see Fig. 1.3). The lowest level is the shared data and knowledge repositories where BIMs and intelligent and reusable objects are held, and these objects are reachable from other levels. Inter-enterprise services correspond to the services provided by model servers and catalogue servers. The data will be distributed using universal standards such as Simple Object Access Protocol (SOAP). The project (or the virtual enterprise) will use envisioned services by integrating them into their standard or in-house developed applications and data stores. User interfaces will allow several different representations (i.e. 2D, 3D and 4D) of the integrated project model, and these representations will be available from different platforms.

Figure 1.3 depicts the envisioned architecture according to the ROADCON project. The developer of the IFC standard, BuildingSmart (formerly, International Alliance for Interoperability), currently illustrates the role of BIMs in three dimensions, namely terminology, processes and data (see Fig. 1.4). BuildingSmart indicates that if these three dimensions can be correctly and accurately defined, and information in all these dimensions can be efficiently communicated, the industry will benefit from better integration in construction projects. For this purpose, BuildingSmart developed three international standards: Data Dictionary

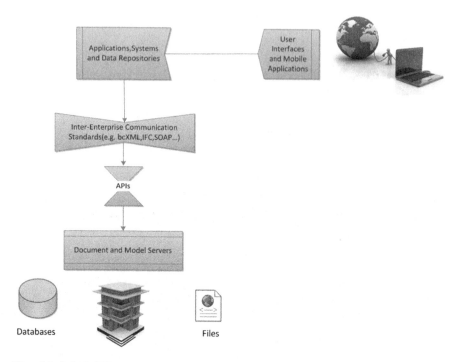

Fig. 1.3 ROADCON—envisioned architecture (images are courtesy of cool design at FreeDigitalPhotos.net)

Fig. 1.4 BuildingSmart triangle

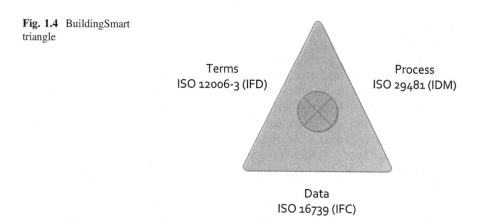

Standard—International Framework for Dictionaries, Process Definition Standard —Information Delivery Manual and Data Modelling standard IFC.

There are also supporting tools such as model view definitions (MVD) which provide definitions of views for specific aspects such as architectural design, structural analysis and HVAC design and so on. Recently, BuildingSmart adopted the Open BIM Collaboration Format (BCF) as a BuildingSmart standard which supports workflow communication in BIM processes.

This chapter provided a brief summary of the history of building information modelling along with the current state of the art. The following chapter will focus on the new era of BIMs and the envisioned role of BIMs for the near future.

References

Eastman, C.: Building Product Models: Computer Environments Supporting Design and Construction. CRC Press, USA (1999)

Eastman, C., Bond, A.H., Chase, S.C.: Application and evaluation of an engineering data model. Res. Eng. Des. **2**(4), 185–208 (1991)

Eastman, C., Wang, F., You, S.F., Yang, D.: Deployment of an AEC industry sector product model. Comput. Aided Des. **37**(12), 1214–1228 (2005)

Gallaher, M.P., O'Connor, A.C., Dettbarn Jr., J.L., Gilday, L.T.: Cost analysis of inadequate interoperability in the U.S. capital facilities industry. NIST Publication GCR 04-867 (2004). Available online at http://www.bfrl.nist.gov/oae/publications/gcrs/04867.pdf

Gielingh, W.F.: General AEC Reference Model GARM, CIB Seminar Conceptual Modelling of Buildings. Lund, Sweden (1988)

Hua, G.B.: A BIM based application to support Cost Feasible 'Green Building' concept decisions. In: Underwoord, J., Isikdag, U. (eds.) Handbook of Research on Building Information Modelling and Construction Informatics: Concepts and Technologies. IGI Global, Hershey (2010)

Kemmerer, S.J.: STEP: the grand experience. NIST Special Publication 939 (1999). Available online at http://www.nist.gov/msidlibrary/doc/stepbook.pdf

Kim, I., Liebich, T.: A data modelling framework and mapping mechanism to incorporate conventional CAD systems into an integrated design environment. Int. J. Constr. Inf. Technol. **7**(2), 17–33 (1999)

Kim, I., Liebich, T., Maver, T.: Managing design data in an integrated CAAD environment: a product model approach. Autom. Constr. **7**(1), 35–53 (1997)

Loffredo, D.: Efficient database implementation of EXPRESS information models. PhD thesis, Rensselaer Polytechnic Institute, Troy, New York (1998). Available online at http://www.steptools.com/~loffredo/papers/expdb_98.pdf

NBIMS.: National BIM standard purpose. US National Institute of Building Sciences Facilities Information Council, BIM Committee (2006). Available online at http://www.nibs.org/BIM/NBIMS_Purpose.pdf

NIST CIS2 Web Site.: Web Site (2005). Available online at http://cic.nist.gov/vrml/cis2.html

Rebolj, D., Babic, N.C., PodBreznik, P.: Automated building process monitoring. In: Underwoord, J., Isikdag, U. (eds.) Handbook of Research on Building Information Modelling and Construction Informatics: Concepts and Technologies, IGI Global, Hershey (2010)

ROADCON Project Deliverable 4.: ICT Requirements of the European Construction Industry: The ROADCON Vision (2001). Available online at http://cic.vtt.fi/projects/roadcon/docs/roadcon_d4_short.pdf

ROADCON Project Deliverable 5.2.: ICT Requirements of the European Construction Industry: The ROADCON Vision (2001). Available online at http://cic.vtt.fi/projects/roadcon/docs/roadcon_d52.pdf

Schenk, A.D., Wilson, R.P.: Information Modelling: The EXPRESS Way. Oxford University Press, New York (1994)

Solis, J.L.F., Mutis, I.: The idealization of an integrated BIM, lean, and green model (BLG). In: Underwoord, J., Isikdag, U. (eds.) Handbook of Research on Building Information Modelling and Construction Informatics: Concepts and Technologies. IGI Global, Hershey (2010)

Spearpoint, M.: Extracting fire engineering simulation data from the IFC. In Underwoord, J., Isikdag, U. (eds.) Handbook of Research on Building Information Modelling and Construction Informatics: Concepts and Technologies. IGI Global, Hershey 2010

Sun, M., Lockley, S.R.: Data exchange system for an integrated building design system. Autom. Constr. **6**(2), 147–155 (1997)

Tolman, F.: Product modelling standards for the building and construction industry: past, present and future. Autom. Constr. **8**(3), 227–235 (1999)

US General Services Administration BIM Guide.: GSA BIM Guide Series 01 (2006). Available online at http://www.gsa.gov/bim

Zamanian, K.M., Pittman, J.H.: A software industry perspective on AEC information models for distributed collaboration. Autom. Constr. **8**(3), 237–248 (1999)

Chapter 2
The Future of Building Information Modelling: BIM 2.0

Abstract The first evolution of BIM was from being a shared warehouse of information to an information management strategy. Now the BIM is evolving from being an information management strategy to being a construction management method. This change in interpretation of BIM is fast and noticeable. Four newly emerging dimensions in management of building information towards transforming BIM to BIM 2.0 focus on enabling an (i) integrated environment of (ii) distributed information which is always (iii) up to date and open for (iv) derivation of new information. The chapter starts with providing recent trends in building information modelling and then elaborates on technologies that will enable BIM 2.0. BIM-based management of the overall construction processes is becoming a major requirement of the construction industry, and the final part of this chapter provides matrices that can be used as a tool for facilitating BIM-based projects and process management.

2.1 Introduction

Over the last half-century ICT has evolved beyond the 'personal' computer to become a strategic asset for business in delivering productivity improvements, extending to provide socio-economic development and growth (European Commission 2006). From the personal computer we have witnessed the emergence of technological advancements such as business systems and applications, visualization, communications, the Internet, mobile/smart/android devices, social networking and most recently, virtualization and cloud computing as part of this revolution. There is no doubt that the effects of the digital age have facilitated considerable changes and improvements to the construction industry and in shaping and modernizing the industry as we currently see in the twenty-first century when compared to the 'traditional', 'archaic' and 'draconian' one from the dim and

© The Author(s) 2015
U. Isikdag, *Enhanced Building Information Models*,
SpringerBriefs in Computer Science, DOI 10.1007/978-3-319-21825-0_2

distant past. It is evident that construction organizations are already in the process of looking towards rapidly maturing technology approaches such as virtualization and cloud computing in the provision of cheaper, more flexible and commoditized ICT infrastructure services to directly drive business efficiencies (France et al. 2010). Other industries are demonstrating that combined cost savings up to 35 % can be achieved through a range of modernization measures, including the consolidation of data centres and full utilization of virtualization technologies. The advancements in information technologies have changed the way we interpret design, analysis and construction management and facilities management. As it has been discussed in the previous chapter, building information modelling has emerged as a cure to the illness of poor interoperability in the industry, but now the paradigm is accepted as a new method of information management, and a new method of construction management. The optimistic view on building information modelling argues that this methodology will be a "*sine qua non*" in the future of the construction industry. The first evolution of BIM was from being a shared warehouse of information to an information management strategy. Now BIM is evolving from being an information management strategy to being a construction management method. The evolution of BIM is fast and noticeable. Today, BIM-based management of daily processes in construction is becoming a de facto standard for large investments. Many projects in the US, Singapore, Dubai and the UK require the existence of BIM-based processes, and involvement of BIM managers (a profession emerged in the past 10 years).The evolution of BIM from BIM 1.0 to BIM 2.0 can be investigated in two dimensions. The first dimension is the changing role of information models (i.e. the shared information sources) from being a shared database to something more complicated. The second dimension is the emerging role of BIM as a new construction management method. We elaborate on these two dimensions in the remainder of this chapter, but it would be good to look at the research dimensions of BIM in the past 5 years as a starting point.

2.2   Research Dimensions of Building Information Modelling

In 2010, Underwood and Isikdag edited *The Handbook of Research on Building Information Modelling and Construction Informatics: Concepts and Technologies*, which was the first collaborative academic book that involved authors from all over the globe.

The handbook (Underwood and Isikdag 2010a, b) contained 28 chapters, which had been the motivation behind defining a research compass for BIM (see Fig. 2.1). The research compass provides the current research directions for BIM. These directions were explained in detail in Isikdag and Underwood (2010), the following will summarize these directions.

Fig. 2.1 BIM research
compass

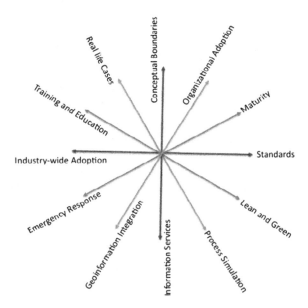

2.2.1   Information Model-Related Aspects

Conceptual Boundaries of the Information Model BIMs have mainly emerged in the form of schema standards for information exchange, i.e. as enablers of data level interoperability. Van Nederveen (Van Nederveen et al. 2010) indicated that a clarification between (a) what is being modelled and (b) how these can be modelled needs to be clearly identified. The research in the area focuses on enhancing the methods and languages for information modelling and reasoning-based approaches to BIM.

Standardization As standardization is a key enabler and facilitator of data level interoperability, this area will continue to be a focus of BIM research. For example, Dado et al. (2010) provide an overview of the characteristics of interesting conceptual product approaches such as standardization, minimal model, core model, NOT, vocabulary and ontology product modelling. In recent years, the US and UK have taken the initiative for developing nationwide BIM standards, which covers more than representation and exchange of information (Suermann and Issa 2010). Such approaches can help in formalizing the information exchanges, the processes and workflows, and will contribute to the evolution of BIM to a project management method.

2.2.2 Organizational Aspects

Adoption The move from CAD-based thinking to the vision of BIM is much more difficult as it involves a shift in fundamental data management philosophy. As indicated by Bew and Underwood (2010), in a similar manner to the move from old accounting packages to Enterprise Resource Planning (ERP), this transformation includes the formal management of processes on a consistent, repeatable basis. Like the ERP implementation, this too is a very difficult transition to make. The lack of mature process management tools and methodologies for the projects has made this transition more confusing. BIM adoption most likely occurs in phases (i.e. in an almost Darwinian evolutionary way), but serious effort should be taken to move from one phase into another.

Maturity A key area in BIM is organizational readiness. If BIM is considered as a set of new technology and methodologies supporting information management in the construction industry, then maturity in terms of implementing and using BIM (technology and methodologies) is critical to the success of BIM implementation. Frameworks for measuring BIM maturity can greatly facilitate organizations in positioning themselves against their competitors in terms of technological, methodological and process maturity. Such a maturity framework is explained in Succar (2010).

Education and Training Education and training is *sine qua non* for successful BIM implementation and adoption in addressing issues/barriers such as culture, etc. As mentioned in Tanyer (2010) and as appeared in the Integrated Project Delivery (IPD) efforts in the US, AEC professionals are beginning to move away from the traditional way of design and project delivery towards a more integrated one. Project-based collaborative learning environments such as in Stanford University (PBLLab 2007), and e-learning environments such as ITC-Euromaster (2009) will also facilitate (and be facilitated by) the use of BIM and collaborative design approaches.

Real-life Cases BIM is not a subject of pure (laboratory type) research any more. This has significantly evolved over the last few years with the implementation of BIM methods and shared digital models in real-life projects increasing exponentially (Lostuvali et al. 2010; Underwood and Isikdag 2010a, b; Riese 2010). The experience and lessons learned from real-life cases will contribute to the development of BIM as a data model or as a project management methodology.

Industry-wide Adoption Research towards the positioning of BIM adoption across disciplines in relation to their current status and future expectations and based on such factors as the tools, people and processes is viewed as a key requirement. For instance Gerrard et al. (2010) provided a bird's-eye view of the industrial adoption picture.

2.2.3 Domain-Specific Aspects

Lean Construction and Green BIM The aim of lean construction is to enable continuous improvement of all construction processes in the building life cycle (starting from design through the demolition of the building) (Solis and Mutis 2010). On the other hand, to address global concerns on environmental issues, the construction industry now takes the initiative to build more 'environment-friendly' buildings, along with reducing its own carbon footprint such as during the construction stage. BIM emerges here as a strong tool where green design, green construction and lean construction can be enabled by the utilization of BIM-based design, simulation and information management tools and methods.

Building and Geo-information Integration As mentioned by Peters (2010), Van Oosterom et al. (2006) and Isikdag and Zlatanova (2008) there is an apparent need for integrated geometric models and harmonized semantics between BIM and geo-information modes for efficient city management.

Emergency Response Emergency response operations indoors require a high amount of geometric, semantic and state information related to the building elements. Until very recently egress models used in building evacuation have mainly been based on 2D floor plans. Today BIM is capable of providing detailed geometric and semantic information related to the buildings, where floor plans, navigation graphs and indoor positioning methods are developed using this information.

2.2.4 Project Management Aspects

Process Simulation and Monitoring The efforts in the area of 4D CAD are making much use of 3D CAD models, but in recent years BIMs have superseded 3D CAD models in the visual simulation of construction processes. Analysis such as clash detection can now be completed using BIM software. BIMs are also used in monitoring the construction progress and as Rebolj et al. (2010) describes, activity progress can be monitored directly by using a combination of data collection methods which are based on the BIM, especially on the 4D model of the building.

2.2.5 Integration and Interoperability Aspects

Building Information Services (BIS) A current trend in the software industry is towards enabling interoperability over Web services. In fact the AEC industry is still not fully benefiting from the service-oriented approaches as the focus of our industry is still quite data integration-oriented. The use of BIM servers is now increasing with open-source implementations such as BIMServer (2010). As explained by London et al. (2010), future BIM approaches would require the shared models in model servers to be linked with external systems in a heterogeneous environment.

2.3 BIM-M: Utilization of BIM in Construction Management

Building Information Modelling and Management (BIM-M) can be defined as the information management process and strategy which covers the whole life cycle of a building (from conception to demolition) and mainly focuses on enabling and facilitating the integrated way of project flow and delivery through the collaborative use of semantically rich 3D digital building models in all stages of the project and building life cycle (Underwood and Isikdag 2010a, b). BIM-M is a model-oriented information management strategy that benefits from the use of building information models. BIM-M is offering significant opportunities in revolutionizing the sector by enabling seamless processes that support the complete life cycle of the facility, embedding a model-based approach and by supporting full information coordination and management. While BIM-M has been in existence in one carnation or another for over 30 years, it is only within the last decade that BIM-M has really begun to receive serious attention particularly from industry and also at a governmental level, which is continuing to gather momentum. BIM-M is currently being employed across the globe on a variety of projects from housing to the prestige, at varied levels of adoption, and within various types of organizations from prime contracting and large consulting organizations to small architectural practices. Clients are now also becoming aware of the potential for BIM-M at the pre- and post-occupancy stage in delivering real value. More recently, government clients across the globe including the US, Denmark, Finland and the UK have begun to implement national initiatives/strategies as a statement of intent in driving BIM-M forward through the procurement of public projects towards establishing industry-wide adoption, which is further contributing to progressing the modernization of the industry (Bew and Underwood 2010). The philosophy that lies behind BIM-M stemmed from four dimensions in relation to the management of building information and these have been agreed by the industry over the last two decades. These dimensions can be summarized as enabling (i) model-based management of (ii) shared building information, which provides (iii) meaningful data about a building/facility in a (iv) standardized way. The first dimension is related to representation of the information within an information model schema which will help the representation of the information in a structured manner. The building information model represents the building elements within an agreed spatial hierarchy and well-defined semantics. The second dimension is related with utilizing the shared way of information management instead of simple exchange of files which causes consistency problems due to versioning. The third dimension is related with utilization of a data model that is based on agreed taxonomies. The final dimension deals with providing the data model as an internationally agreed standard, for which many types of software would be able to develop input and output plug-ins to generate and read the contents of the exchanged model file; in addition this enables databases to provide application programming interfaces for interacting with the standard information model.

2.4   Technologies for BIM 2.0

In last 5 years, development of new technologies and approaches in information management provides interesting opportunities for BIM-based information management. Underwood and Isikdag (2011) stated that four newly emerging dimensions in management of building information towards transforming from BIM to BIM 2.0 focus on enabling an (i) integrated environment of (ii) distributed information which is always (iii) up to date and open for (iv) derivation of new information. Various information technologies can facilitate this new focus such as cloud computing, sensor networks, stateless Web services and semantic Web. The following elaborates on the role of these technologies in making BIM 2.0 a reality.

Cloud Computing   The term cloud computing indicates the use of Internet (i.e. the cloud) for managing highly scalable and customizable virtual hardware and software resources (which are provided as services). Cloud computing today is broken down into three segments as providing "Software as a Service (SaaS)", "Platform as a Service (PaaS)" and "Infrastructure as a Service (IaaS)". Cloud computing also involves the virtualization of storage environments to form a virtual data centre which is available over the Internet. The construction industry will benefit from cloud computing mostly by making use of SaaS approach and data centre virtualization. Applications used in various stages of the building's life cycle will work within a distributed environment (i.e. offered as software services), and the information backbone of the construction project or building (i.e. BIMs) will reside in a virtual data centre (and offered as a data service).

Sensor Networks   Recent developments in the field of BIM-M have shown that BIMs are very successful in presenting sematic information about building elements along with their geometric representation. Although the information in BIMs is meaningful, it in fact becomes stateless after the construction of the building is completed. In other words, a BIM user can find out whether a door in a building is constructed of timber, or the door has been constructed (or not) on a given date, but it is not possible to get informed on whether that door is open or closed at a certain point in time only by using BIMs or BIM-based information infrastructures. In this situation, up-to-date building information will be provided by sensors, or by a network of sensors which are monitoring the building. In the context of BIM-M, the distributed sensors and sensor networks, Internet of Things (IoT) nodes will monitor conditions such as temperature, gas levels, pollutants, humidity, state of doors and windows (i.e. being open/closed and so on), occupancies in rooms and conditions of different systems working within a building/facility. These issues will be elaborated in more detail in the following chapters of this book, in the context of IoT.

RESTful Web Services Service-oriented architectures and RESTful Web services offer opportunities for making building information stateful (i.e. real time, accurate and up to date). Vast amounts of residing in BIMs, and micro (atomic) feeds from sensor networks can be exposed as loosely coupled Web services, where generating data mashups from these resources would be very straightforward.

Semantic Web If this mass of new information (derived from multiple resources) can be restructured in compliance with semantic Web Standards and supported by well-built ontologies, i.e. formal specifications of conceptualizations which consist of finite list of terms and the relationships between these terms (Antoniou and van Harmelen 2008), semantic queries such as "Would you provide me the number of working elevators and escalators in the Empire State Building between 12:00 and 14:00?", "Would you provide me the average CO_2 level in top 20 floors of 5 of the highest buildings in London?" or "Please provide me the difference between temperatures in my hotel room in Singapore Marina Bay, and my office in Sydney" can be responded. The success rate in responses to the (presented) semantic queries will depend on, (i) the level of integration of distributed building information, (ii) the level of success in derivation of information mass from multiple loosely coupled resources (which are exposed as Web services), and finally (iii) how well the query can be interpreted and, reasoning/search and retrieval can be accomplished upon the interpreted query.

2.5   BIM-Based Management of Construction Processes

The developments in the field of building information modelling have led the stakeholders in the construction projects to re-engineer their traditional construction management processes to BIM-based design and construction processes. Although there does not exist an ISO standard for management of BIM-based processes, a joint effort (known as BIM Project Execution Planning) in the last 5 years has produced noticeable scientific outputs. As stated in the project website (BPEPG 2015) the planning guide which is produced as a result of this research aimed to provide a practical manual that can be used by project teams for designing their BIM strategy and developing a BIM Project Execution Plan. This guide provides a structured procedure for creating a BIM Project Execution Plan. The four steps within the procedure include:

1. Defining high-value BIM uses during project planning, design, construction and operational phases
2. Using process maps to design BIM execution
3. Defining the BIM deliverables in the form of information exchanges

4. Developing a detailed plan to support the execution process

The project is started by defining BIM uses, defined as "a method of applying Building Information Modelling during a facility's life cycle to achieve one or more specific objectives". The uses can be interpreted as use cases that cover multiple processes of BIM-based information management. As explained in PSU (2013), the defined BIM uses covered the overall life cycle of the project from inception to finalization. The project also elaborated on the concept of level of development (LOD) stating that for each of the BIM uses, the LOD should be identified in order to maximize the benefit from the BIM use. The LOD describes the level of detail/granularity to which a model element is developed. The LOD specification for BIM is originally developed by BIM Forum (2013) and presents six levels of development (Table 2.1).

The project presented 25 BIM uses as:

1. Existing conditions modelling
2. Cost estimation
3. Phase planning
4. Programming
5. Site analysis
6. Design reviews
7. Design authoring
8. Structural analysis
9. Lighting analysis
10. Energy analysis

Table 2.1 BIM levels of development (BIM Forum 2013)

LOD 100	The model element may be graphically represented in the model with a symbol or other generic representation, but does not satisfy the requirements for LOD 200. Information related to the model element (i.e. cost per square foot, tonnage of HVAC, etc.) can be derived from other model elements
LOD 200	The model element is graphically represented within the model as a generic system, object or assembly with approximate quantities, size, shape, location and orientation. Non-graphic information may also be attached to the model element
LOD 300	The model element is graphically represented within the model as a specific system, object or assembly in terms of quantity, size, shape, location and orientation. Non-graphic information may also be attached to the model element
LOD 350	The model element is graphically represented within the model as a specific system, object or assembly in terms of quantity, size, shape, orientation and interfaces with other building systems. Non-graphic information may also be attached to the model element
LOD 400	The model element is graphically represented within the model as a specific system, object or assembly in terms of size, shape, location, quantity and orientation with detailing, fabrication, assembly and installation information. Non-graphic information may also be attached to the model element
LOD 500	The model element is a field-verified representation in terms of size, shape, location, quantity and orientation. Non-graphic information may also be attached to the model elements

11. Mechanical analysis
12. Other engineering analysis
13. LEED evaluation
14. Code validation
15. 3D coordination
16. Site utilization planning
17. Construction system design
18. Digital fabrication
19. 3D control and planning
20. Record model
21. Maintenance scheduling
22. Building system analysis
23. Asset management
24. Space management/tracking
25. Disaster planning

Based on the BIM uses and BIM LODs, BIM-based processes can be managed. Once the BIM uses for the projects and BIM LODs that will be used in each BIM Use are determined, the BIM Execution Planning Matrices need to be developed in order to manage the BIM-based processes. Examples of such matrices are developed by the author, and provided below.

BIM based process management matrix

Activity	Actors/role	Model/view owner	Software in use	Inputs	Outputs	BIM use	BIM LOD	BIM objects required	Exchange formats
						.			

Actor/Role matrix

Activity	Actor	Role	Sub-role	Eligibility of model ownership

Phase/Activity/LOD matrix

Project phase	Activity	BIM LOD	BIM objects required

BIM USE/LOD matrix

BIM use	BIM LOD	BIM objects required

As also stated in PSU (2013) there is no single best method for BIM implementation on every project, each team must effectively design a tailored execution strategy by understanding the project goals, project characteristics and the capabilities of the team members. For the construction industry, there is still a long way to go and much to do in terms of realizing the full potential of these emerging technologies in line with the efficiencies and performance improvement that are being witnessed in other sectors. However, as both the technologies along with the industry (in their capability to embrace them) further mature, and as BIM-based process and information management techniques get more advanced, progressive improvements towards BIM 2.0 will continue to be made, enabled through emerging technologies.

References

Antoniou, G., van Harmelen, F.: A semantic web primer. The MIT Press, Cambridge (2008)

Bew, M., Underwood, J.: Delivering BIM to the UK market. In: Underwoord, J., Isikdag, U. (eds.) Handbook of Research on Building Information Modelling and Construction Informatics: Concepts and Technologies. IGI Global, Hershey (2010)

BIM Server: Open Source BIM server. Available online at: http://www.bimserver.org (2010). Accessed 05 Jan 2010

BIM Forum: BIM LOD specification. Available online at http://bimforum.org/wp-content/uploads/2013/08/2013-LOD-Specification.pdf (2013)

BPEPG: BIM Project Execution Planning Guide Web Site. Available online at http:///bim.psu.edu (2015)

Dado, E., Beheshti, R., Van de Ruitenbeek, M.: Product modelling in the building and construction industry: a history and perspectives. In: Underwoord, J., Isikdag, U. (eds.) Handbook of Research on Building Information Modelling and Construction Informatics: Concepts and Technologies. IGI Global, Hershey (2010)

European Commission: ICT Uptake, Working Group 1. ICT Uptake Working Group draft Outline Report. Available at: http://ec.europa.eu/enterprise/ict/policy/taskforce/wg/wg1_report.pdf (2006)

France, K., Fox, S., Khosrowshahi, F., Underwood, J.: Building on IT Survey: Cost Reduction and Cost Effectiveness. Construct IT For Business Report, Salford (2010)

Gerrard, A., Zuo, J. Zillante, G., Skitmore, M.: Building information modeling in the Australian architecture engineering and construction industry. In: Underwoord, J., Isikdag, U. (eds.) Handbook of Research on Building Information Modelling and Construction Informatics: Concepts and Technologies. IGI Global, Hershey (2010)

Isikdag, U., Zlatanova, S.: Towards defining a framework for automatic generation of buildings in CityGML using building information models. In: Lee, J., Zlatanova, S. (eds.) 3D Geo-Information Sciences, Springer LNG&C, Berlin (2008)

ITC-Euromaster: The European Master in Construction Information Technology. Available online at http://euromaster.itcedu.net/ (2009). Accessed 11 Dec 2009

London, K., Singh, V., Gu, N., Taylor, C., Brankovic, L.: Towards the development of a project decision support framework for adoption of an integrated building information model using a model server. In: Underwoord, J., Isikdag, U. (eds.) Handbook of Research on Building Information Modelling and Construction Informatics: Concepts and Technologies. IGI Global, Hershey (2010)

Lostuvali, B., Love, J., Hazleton, R.: Lean enabled structural information modeling. In: Underwoord, J., Isikdag, U. (eds.) Handbook of Research on Building Information Modelling and Construction Informatics: Concepts and Technologies. IGI Global, Hershey (2010)

Riese, M.: Building lifecycle information management case studies. In: Underwoord, J., Isikdag, U. (eds.) Handbook of Research on Building Information Modelling and Construction Informatics: Concepts and Technologies. IGI Global, Hershey (2010)

PBLLab: Stanford University Project Based Learning Laboratory. Available online at http://pbl. stanford.edu (2007). Accessed 11 Dec 2009

Peters, E.: BIM and geospatial information systems. In: Underwoord, J., Isikdag, U. (eds.) Handbook of Research on Building Information Modelling and Construction Informatics: Concepts and Technologies. IGI Global, Hershey (2010)

PSU: The Uses of BIM: Classifying and Selecting BIM Uses. Available online at http://bim.psu. edu/Uses/the_uses_of_BIM.pdf (2013)

Rebolj, D., Babic, N.C., PodBreznik, P.: Automated building process monitoring. In: Underwoord, J., Isikdag, U. (eds.) Handbook of Research on Building Information Modelling and Construction Informatics: Concepts and Technologies. IGI Global, Hershey (2010)

Solis, J.L.F., Mutis, I.: The idealization of an integrated BIM, lean, and green model (BLG). In: Underwoord, J., Isikdag, U. (eds.) Handbook of Research on Building Information Modelling and Construction Informatics: Concepts and Technologies. IGI Global, Hershey (2010)

Succar, B.: Building information modelling maturity matrix. In: Underwoord, J., Isikdag, U. (eds.) Handbook of Research on Building Information Modelling and Construction Informatics: Concepts and Technologies. IGI Global, Hershey (2010)

Suermann, P.C., Issa, R.R.A.: The US national building information modeling standard. In: Underwoord, J., Isikdag, U. (eds.) Handbook of Research on Building Information Modelling and Construction Informatics: Concepts and Technologies. IGI Global, Hershey (2010)

Tanyer, A.M.: Design and evaluation of an integrated design practice course in the curriculum of architecture. In: Underwoord, J., Isikdag, U. (eds.) Handbook of Research on Building Information Modelling and Construction Informatics: Concepts and Technologies. IGI Global, Hershey (2010)

Underwood, J., Isikdag, U.: A synopsis of the handbook of research on building information modelling. CIB World Congress, Salford, 10–13 May (2010a)

Underwood, J., Isikdag, U.: Handbook of Research on Building Information Modelling and Construction Informatics: Concepts and Technologies. IGI Global, Hershey (2010b)

Underwood, J., Isikdag, U.: Emerging technologies for BIM 2.0. Constr. Innov. 11(3), 252–258 (2011)

Van Nederveen, S., Beheshti, R., Gielingh, W.: Modelling Concepts for BIM. In: Underwoord, J., Isikdag, U. (eds.) Handbook of Research on Building Information Modelling and Construction Informatics: Concepts and Technologies. IGI Global, Hershey (2010)

Van Oosterom, P., van Stotter, J., Janssen, E.: Bridging the worlds of CAD and GIS. In: Zlatanova, Prosperi (eds.) Large-scale 3D data integration -Challenges and Opportunities, pp. 9–36. Taylor&Francis, London

Chapter 3
Foundational SOA Patterns for Complex Information Models

Abstract In domains where detailed semantic information coupled with detailed geometric representations is of key importance, such as city modelling, construction, aircraft industry, ship production and so on, information models that represent these domains are usually of a complex structure. This chapter starts by summarizing design principles of service orientation, and later provides 10 service-oriented architecture (SOA) patterns for managing complex information models. The chapter provides patterns to facilitate the management of complex information models. Generalized service-oriented architectural approaches covering complex information models are presented. BIM-specific SOA patterns are presented in Chaps. 6 and 7.

3.1 Introduction

The term design pattern is used heavily in software engineering. Design patterns prevent developers from re-inventing the wheel when they face a common problem in the software design. A design pattern defines a problem that frequently occurs in software design and implementation, and then describes the solution to the problem in such a way that it can be reused in many different situations. A pattern portrays a commonly recurring structure of communicating components that solve a general design problem within a particular context (Gamma et al. 1995; Yacoub and Ammar 2003). As explained by Isikdag (2012), a pattern can be characterized as the template for a solution, a software design problem or as a defined and recognized formalization of interaction between the software components for fostering better software design. The use of patterns in software design helps software developers in design decisions, (i) when they face commonly observed problems in the design of their new software or (ii) when they would like to introduce new ways of interaction

© The Author(s) 2015

U. Isikdag, *Enhanced Building Information Models*,
SpringerBriefs in Computer Science, DOI 10.1007/978-3-319-21825-0_3

between the different layers or components in their design. Conventional approaches for sharing and exchanging information models in AEC industry, e.g. by using BIMs (as explained in the previous section), are mostly focused on exchanging a physical file between applications and using shared central databases to perform CRUD (create/read/update/delete) operations on these models (i.e. BIMs). Furthermore, today the use of model views (i.e. subsets of BIM defined according to the needs of a specific information exchange scenario) is encouraged for facilitating the BIM-based information management throughout the lifecycle of a building. As indicated in Isikdag (2012), recent R&D efforts such as BIMServer (BIMServer 2015) have shown that it is possible to use Web services for acquiring information from BIMs. Although the literature in the field illustrates examples of the use of shared (central) databases and Web services to obtain information from BIMs, there have been very few efforts (such as Isikdag 2012) on formalizing the (i) sharing of a distributed BIM and (ii) information acquisition from the shared BIM over the Web.

3.2 Design Principles of Service Orientation

He (2003) indicated that two constraints exist for implementing Web services: (i) Interfaces must be based on Internet protocols such as HTTP, FTP and SMTP and (ii) except for binary data attachments, messages must be in XML. This has also changed by the emergence of RESTful architectures. Two definitive characteristics of Web services are loose coupling and network transparency (Pulier and Taylor 2006). The foundational layers for the design principles of service-oriented architectures (SOA) are defined in Gamma et al. (1995), Hohpe and Woolf (2003), Linthicum (2003) and Fowler (2003). In his well-known textbook, Erl (2009) defined SOA design patterns based on nine design principles explained in Erl (2008), which will be summarized in this section.

Standardized Service Contract A service contract can be defined as a set of rules that defines a service. These rules are represented in a data model for the Web service (if such a data model exists). The contract can be interpreted as the meta-model of the Web service. The model is used to express the purpose and capabilities of the service. As indicated in Erl (2008), the standardized service contract design principle is perhaps the most fundamental part of service orientation, in that it essentially requires that specific considerations be taken into account when designing a service's public technical interface and assessing the nature and quantity of content that will be published as part of a service's official contract. In Simple Object Access Protocol (SOAP) architectures, the contract is explicit and represented in the form of WSDL (XML) files, but in RESTful architectures (which are much popular today) there is no explicit service contract that is represented with an XML schema and file, but for service discovery purposes it is recommended to prepare such a file or documentation for facilitating the discovery of the service.

Loose Coupling Coupling refers to a connection or relationship between two things. A measure of coupling is comparable to a level of dependency. As explained by Pulier and Taylor (2006), in a traditional distributed environment, computers are tightly coupled, i.e. each computer connects with others in the distributed environment through a combination of proprietary interfaces and network protocols. Web services in contrast are loosely coupled, i.e. when a piece of software has been exposed as a Web service, it is relatively simple to move it to other hardware. In addition, this design principle advocates that coupling between different Web services should also be loose. So we need to consider two loose coupling principles, i.e. services should not be dependent on the existence of a specific hardware, and services should not be dependent on the existence of another service. Loose coupling principle promotes the independent design and evolution of a service's logic and implementation while still guaranteeing the baseline interoperability with consumers who have come to rely on the service's capabilities (Erl 2008).

Service Abstraction In the terminology of object-oriented modelling, abstraction is closely related with information hiding and refers to hiding the unnecessary details of objects. Similarly, as indicated by (Erl 2008), in service-oriented thinking this principle emphasizes the need to hide as much of the underlying details of a service as possible. The principle helps to preserve the loosely coupled relationship between the Web services. Abstraction also plays an important role in building up and configuring the service compositions.

Service Reuse Reusability principle indicates that once composed, a piece of code can be used in many applications with minimal change. Once a code is used in a new application, the new application inherits and implements the methods and attributes that are defined by this code. Depending on the quality of the code, the quality of the application may also increase. Similarly, Web services need to be designed in compliance with the reusability principle; and by this, the individual core services would become reusable Web components that can be used as lego bricks in the construction of more complicated Web services. The core services can reside inside the service containers or service pools where many large-scale services can make use of these services as resources.

Service Autonomy As indicated by Erl (2008), for services to carry out their capabilities consistently and reliably, their underlying solution logic needs to have a significant degree of control over its environment and resources. Service autonomy advocates that the services should be provided with improved independence from their execution environments. This provides more flexibility and reliability when services operate with less dependence on resources of the environment that they are running.

Service Statelessness In n-tier software architectures, the server side is responsible for keeping the state of each client or application. This responsibility causes the server to consume a lot of hardware resources and also the Web service becomes very slow. In order to solve this problem, some middleware components are introduced. In fact, in an SOA where reusability is heavily utilized, managing the state data becomes more and more resource consuming either for server or for middle-tier components. The service statelessness principle defends that the Web services should not store and manage the state information but this information needs to be managed by other external components that are specifically designed for these purposes. Services should only be designed as stateful when this is a real architectural requirement. This principle forms the foundation of RESTful services.

Service Discoverability   Discoverability is defined as the ability of something to be found. In our case, service discoverability refers to a service's ability to be discovered by another service or application. Discoverability is directly proportional to accessibility of the service and also contributes to the usability of the service. Discoverability of Web services can be enabled and facilitated by use of global service registries such as UDDI. In fact, such a global registry of Web services is not sufficient; it is also very important to generate services that are self-discoverable. Meta-information about the service purpose and characteristics needs to be embedded into Web services that will help search engines to discover the service. Sites such as www.apis.io or www.programmableweb.com are well-known examples of service search engines and can be used to discover Web services defined for any purpose.

Service Composability   In computer science, object composition refers to the ability to combine core or simple objects into more complex objects. In composition, paradigm simple objects act as building blocks of more complex objects. A similar approach can be applied in SOAs to generate composite services by using or reusing core Web services as (service) components. To achieve this, Web services need to be defined in a way that they are pluggable to other services.

Service Interoperability   Interoperability of software can be defined as the capability of one type of software to function as the other, or as the capability of different software to exchange and share data via a common set of exchange formats. The latter is known as data-level interoperability. Interoperability is also possible at the service level when one service is capable of substituting the function of another Web service, or when one Web service is able to operate using data models that are consumed by the other Web service. Hence in order to be considered as interoperable services, they should either have the ability to function as the other, or have the ability to operate with the other services' data model. Interoperability of services is a key element in facilitating the integration of services, where one-to-n interaction between services is a common practice.

3.3   Complex Information Models

It is not easy to define what a complex information model is, but we can start from simple information models and try to underline the difference between simple and non-simple (complex) information models. Information models for most domains,

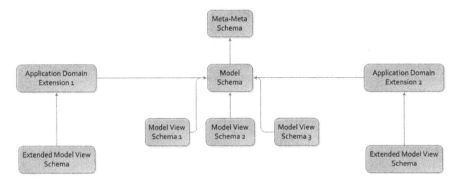

Fig. 3.1 A generalized view of complex information models

ranging from health care to plane ticket reservation systems make use of a single schema that is defined for that specific domain (or application purpose). Once the model is agreed on by the software developers, it is used to generate tables in a shared database. Once there appears a need, different views of the model are then generated using the database management system and stored inside the database. The schemas of the information model and model views are stored and exchanged very rarely outside the database management system environment. The information management practice explained here is the common practice for the daily routines in most domains in production and service industries. In this common practice, schema exchange and data-level interoperability are not common requirements. On the other hand, in domains where detailed semantic information coupled with detailed geometric representations is of key importance (such as city modelling, construction, ship and aircraft production and so on), information models become complex. In these domains, information models are represented by schemas which are agreed by the major stakeholders of the domain (also defined in the form of standards). The model schemas usually refer to a meta-model or data modelling standard (such as ISO 10303 EXPRESS or XML). The model schema forms the core of the information model standards (such as IFC or CityGML schemas). Model views are also represented by agreed schemas. Furthermore, some of these models such as CityGML (OGC 2012) are capable of being extended using application domain extensions (ADEs) which can be referred as extension schemas. In some cases such as models defined in ISO 10303, the model definition language (such as EXPRESS) needs to be recognized by the databases; thus, a meta–meta schema (i.e. a schema that defines the data definition language (DDL), i.e. ISO 10303 EXPRESS) would also be required by the database. As all of the well-recognized databases are capable of interpreting the XML schemas, there would not be such a (meta–meta schema) need for XML compliant information models. Depending on these, a generalized definition of complex information model would contain four components (see Fig. 3.1).

1. Meta–meta schema: Schema of DDL
2. Model Schema: Definitions of the entities/objects of the core model
3. Model-View Schema(s): Definitions of the entities of the sub-model(s)
4. Application Domain Extension Schema(s): Definition of the application-specific entities that are not present in the core model.

3.4 Service-Oriented Patterns

This section elaborates on patterns defined for enabling SOAs for systems using complex information models in their data layer. Patterns defined in this section focus on the provision and facilitation of Web services for complex information models. In some patterns, there are observer-type services. Observer-type Web services explained in this section are composite Web services, but illustrated as a single component in the figures for preventing confusion for the readers. The reader should note that the observer services explained are not simple services that can only be invoked by HTTP requests; in fact, they would be interpreted as a set of software components including an HTTP Web service. The observer services would consist of software components required to observe the changes in the models and present these changes to the other software components upon a request or as a result of subscription. For further information on database and file change detection which are key research topics on their own, readers are advised to check the related computer science literature.

3.4.1 Data Definition Language Provider

Some data modelling languages such as ISO10303 EXPRESS need to be introduced to the model server databases. Once the DDL is introduced, the database would then become capable of understanding the schemas defined using this DDL. The schema that introduces the DDL to the database is known as the meta–meta schema. Once this meta–meta schema is introduced to the database, the database can interpret the meta schemas/model schemas (i.e. the ones that define the information model such as IFC model schema for BIM). The object model DDL can be served through a DDL provider service, which will serve the full schema of the DDL. A service consumer in this case can be an API that has the capability of updating the database or an application that has the capability to update the database. The database would then be updated by the service consumer, and having imported the object model of the DDL, the database would become ready for the data model schema import and recognition. EPMTech's EDM model server (2015) provides an example of such an object database where this pattern can be implemented (Fig. 3.2).

Fig. 3.2 Data definition
language provider

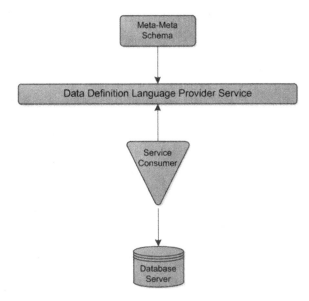

3.4.2   Model View Selector

As mentioned in Chap. 1, model views are facilitators for exchange of information
with software working in specific sub-domains (for instance architectural design,
structural analysis, HVAC design and electrical design are sub-domains of building
design). Depending on the domain requirements, a model can have multiple model
views (or model subsets), which are generated for the specific applications in these
domains. Once the entities of the core model and model views are generated, the
business processes might require working with multiple model views simultaneously. In such a case, several applications of sub-domains would make requests to
the model views (as a whole dataset). Once these requests are made, a model view
gateway service would present the requested model view as the service output. The
service can be consumed by a service consumer API. The API can be embedded
within an application, or the output of the API can be served, or visualized through
a Web interface. The implementation of the pattern does not require the existence of
any specific environment such as the model or view container, but it is assumed that
the model views are persistent in a file system or in a database (Fig. 3.3).

Fig. 3.3 Model view selector

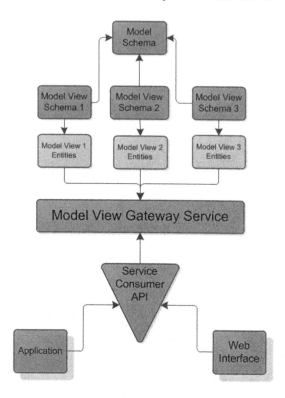

3.4.3   *Model View Entity Extractor*

The idea and background of this pattern is similar to the model view selector, as the pattern works within an environment where there are multiple model views. In the previously presented pattern, model views are presented as a whole dataset depending on the selection of the consuming application. For instance, a design application might require a design view of a BIM (where all entities related to design will be transferred to the client), while an analysis application would require an analysis view of the BIM (where all entities related to analysis will be transferred to the client). In fact in some cases, the model views might become too large and applications sometimes require individual entities of the view. An example of this would be a design application information request made while an architect is working on the detailed design of a single façade element (such as a logo of a company that will be placed in the building façade). In this situation, transfer of the whole dataset of the view (i.e. all façade elements) is unnecessary and will cause redundant use of hardware and network resources. The view entity extractor service defined in this pattern would respond to the request of transferring an individual entity of the view, to the service consumer. Similar to the previous pattern, the service can be consumed by a service consumer API. The API can be embedded

Fig. 3.4 Model view entity
extractor

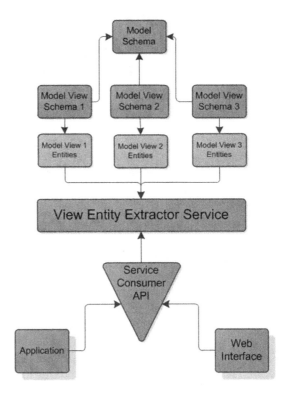

within an application, or the output of the API can be served, or visualized through a Web interface (Fig. 3.4).

3.4.4 Sub-view Generator

The structure of this pattern is similar to the model view selector and model view entity extractor as the pattern works within an environment where there are multiple model views. In the previous pattern, the model view entity extractor service is capable of presenting each individual entity of the model (such as presenting a door, a window, a wall, etc., as a result of a request from a BIM model view). In fact in some cases, business processes require a set of entities of a sub-model of the view (i.e. a sub-view) to be presented as a result of a request. For instance, during the construction of buildings schedules are updated once the construction is progressing. In fact, the update is done as the work progresses in a part of the building (e.g. beams of floor 2 might be constructed on 2 October, and this information needs to be updated on the BIM model view—that is related to scheduling). In this situation, a sub-view would be required to be transferred to the client. As these operations are usually done using mobile devices, the transfer of sub-view would

Fig. 3.5 Sub-view generator

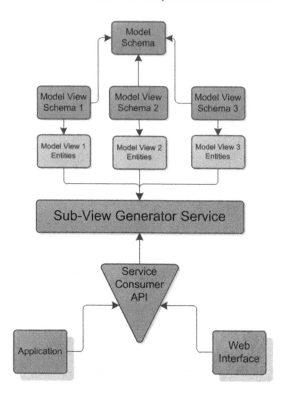

also increase the efficiency of the system. Depending on the user request, the sub-view generator service will generate a sub-view from the model view and present this view as the service output. Similar to the previous pattern, the service can be consumed by a service consumer API. The API can be embedded within an application, or the output of the API can be served, or visualized through a Web interface (Fig. 3.5).

3.4.5   View Observer

The view observer pattern is defined for an environment where there are multiple model views. The pattern is proposing a mediator between the model and the view (s). The previous three patterns are related to presenting model view(s) to the client applications through a Web service and a consumer API. Once these applications work on the model views, they will update the model views inevitably. If we think of a situation where there are 3–4 applications working simultaneously with 3–4 different views and updating them synchronously, this might bring inconstancies between the model views. For example, an architect might be working on the design of a space in a room on model view A, while an engineer might propose a

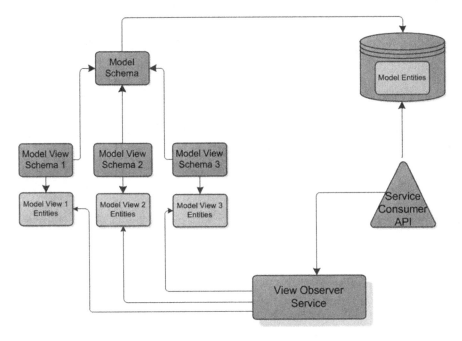

Fig. 3.6 View observer

curtain wall dividing this space into two subspaces on model view B. In this case, the architect needs to be notified immediately regarding the proposed curtain wall. In order to enable this, every change occurring in the model views needs to be listened (or observed) and the model itself (where views are generated) needs to be updated. The view observer service in this pattern is a composite service that includes components for noticing the changes in the model views. The service components listen for changes in multiple model views and once a change occurs in one of the views, it will issue a notification with the details of the change. These notifications would then be consumed by a service consumer API (an observer of the view observer), which is also capable of updating the information model residing in a database (or in a file—which is a less common situation). Although not provided in the diagram, the existence of a database API or file I/O API can be required in order to update the database or the model file (Fig. 3.6).

3.4.6   View Updater

The pattern can be considered as a complementing pattern to the previous pattern view observer. Once the view observer service and the service consumer API update the model, the (core) model becomes the most accurate and the most up-to date resource of information. In fact, it is also key to update the views based on the

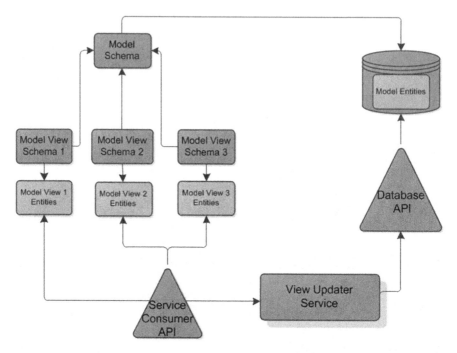

Fig. 3.7 View updater

latest version of the model. This is accomplished using the view updater service.
The view updater will be designed as a composite service (which includes components for) functioning as an observer of the core model (also by maintaining a log
of changes on the DB side). Once a change in the core model is made and is notified
by the view updater service, the service would then indicate which elements from
the core model need to be transferred to update each model view. The information
from the core model will be acquired through a database API that is interacting with
the model server database where the core model resides. A service consumer API
will be used to consume information from the view updater service and update the
model views related to the changes in the core model (Fig. 3.7).

3.4.7   Extended Model Observer

As mentioned earlier, model extensions are defined by application domain extension schema(s) which are used to describe the application-specific entities that are
not present in the core model (Fig. 3.8).

These entities, for instance, can represent specific elements of the urban furniture
for a city model. In this case, the core city model is not required to have these
furniture definitions in its core schema, an extension (can be thought as a data

Fig. 3.8 Extended model
observer

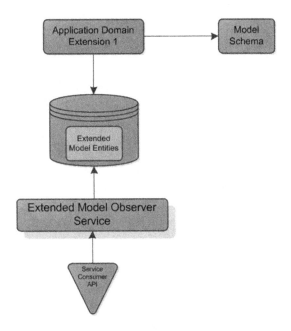

interface) when implemented makes the model capable of representing furniture
entities. The extended model would then be able to represent all objects in the core
model and objects introduced by the application domain extension schema. Novel
developments in the field of Internet of Things (IoT) introduced new real-time
monitoring paradigms where every real-life object can publish information and be
monitored as a part of their machine-to-machine (M2M) communication. In order to
represent feeds coming from objects, an application domain extension can be
introduced. In this case, the resulting extended model will represent real-time
information published by the objects. An example of this can be a city model
extension that is targeted to provide information on real-time traffic density on
roads. Such information will be stored in real-time within the extended city model
entities. In order to notice the changes in the extended model and inform users
about the changes, an extended model observer service needs to be implemented.
This service (is a composite service containing components and) would act as a
listener/observer to the extended model. Once a change occurs, the service will
notify its subscribers. Being one of the service subscribers, a service consumer API
would be beneficial for serving this information to requesting parties.

3.4.8 Extended Model View Observer

This pattern is similar to the previously explained extended model observer pattern.
In this pattern, the extended model view observer service observes the view(s) of

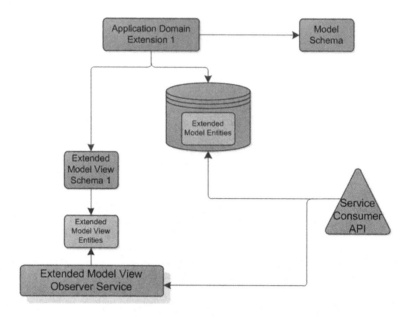

Fig. 3.9 Extended model view observer

the extended model (is a composite service that includes components for noticing the changes in the extended model view). Once the extended model view is handled by an application, changes are expected to occur in this view. At the time a change occurs at the extended model view, the extended model view observer service will notify its subscribers about this change. The most obvious subscriber of this service would be a service consumer API. The main responsibility of the API in this case would be updating the extended model once a change occurs in the extended model view. The role of the API is exactly the same as the role of the API in view observer pattern. In the view observer pattern, it was mentioned that the service consumer API would be able to update the database where the model resides, or would be able to update the model file. The situation is the same for the API in this pattern. Although not provided in the diagram, the existence of a database API or file I/O API can be required in order to update the database or model file (Fig. 3.9).

3.4.9   Extended Model View Updater

This pattern has common elements with the view updater pattern. It is mentioned in the extended model view observer pattern that once a change occurs in one extended model view, the extended model will get updated. In this situation, the extended model becomes the most up to date and accurate model in the system. In fact, other extended model view(s) are not aware of this change, and cannot be

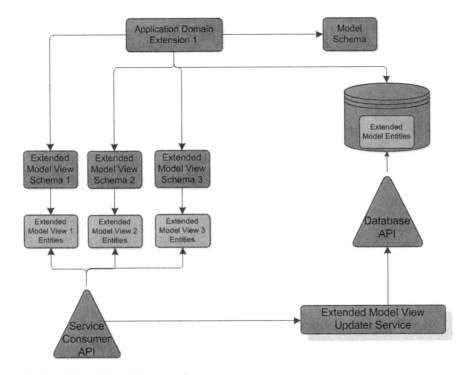

Fig. 3.10 Extended model view updater

considered as accurate views. The extended model view updater service will be designed as a composite service (which include components for) functioning as an observer of the core model (also by maintaining a log of changes on the DB side). Once a change occurs in the extended model, the extended model view updater service would take the responsibility to broadcast the changes with the help of a database API and service consumer API. This latter API will be used to consume information from the extended model view updater service and update the extended model views regarding the changes occurred (Fig. 3.10).

3.4.10   Model Controller

The previous patterns in this section provide observer and updater functions with different services. In fact, these services can be merged into a single service which is known as a controller. The controller in the well-known design pattern MVC covers both functions of observer and updater. Similarly, a Web service can mimic the functionality of the controller for complex information models. In this pattern, a model controller service utilizes a database API (i) in order to listen and broadcast changes in the model and (ii) to update the model based on update requests coming

Fig. 3.11 Model controller

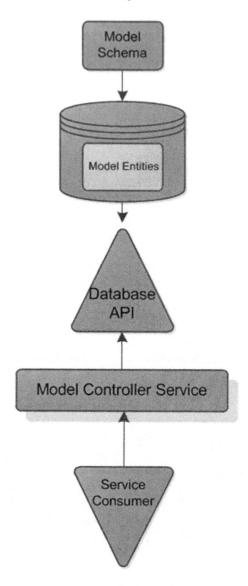

from service consumer. Service consumer, like model controller service, has two functions: to broadcast the changes in the information model and to trigger the update of the model using the model controller service. The pattern can also be generalized to cover model views and extended models and extended model views, for generating model view controller (but should not be mixed with well-known M-V-C pattern), extended model controller and extended model view controller patterns (Fig. 3.11).

References

BIMServer.: BIM Server Web Site (2015). Available online at www.bimserver.org

EDM Model Server.: Software Product (2015). http://www.epmtech.jotne.com/express-data-manager-edm

Erl, T.: SOA: Principles of Service Design. Prentice Hall (2008)

Erl, T.: SOA Design Patterns. Prentice Hall (2009)

Fowler, M.: Patterns of Enterprise Application Architecture. Addison-Wesley (2003)

Gamma, E., Helm, R., Johnson, R., Vlissides, J.: Design Patterns: Elements of Reusable Object-Oriented Software. Addison-Wesley, Reading, Mass. (1995)

He, H.: What Is Service-Oriented Architecture. (2003) http://webservices.xml.com/pub/a/ws/2003/09/30/soa.html. Accessed 21 July 2014

Hohpe, G., Woolf, B.: Enterprise Integration Patterns: Designing, Building, and Deploying Messaging Solutions. Addison-Wesley, Indianapolis (2003)

Isikdag, U.: Design patterns for BIM-based service-oriented architectures. Autom. Constr. **25**, 59–71 (2012)

Linthicum, D.S.: Next Generation Application Integration: From Simple Information to Web Services. Addison-Wesley, Indianapolis (2003)

OGC.: OGC City Geography Markup Language (CityGML) Encoding Standard (2012). Available online http://www.opengeospatial.org/standards/citygml

Pulier, E., Taylor, H.: Understanding Enterprise SOA. Manning Publications, Greenwich, USA (2006)

Yacoub, S.M., Ammar, H.H.: Pattern Oriented Analysis and Design. Addison-Wesley (2003)

Chapter 4
Internet of Things: Single-Board Computers

Abstract The present can be regarded as the start of the Internet of Things (IoT) era. IoT covers the utilization of sensors and near-field communication hardware such as RFID or NFC, together with embedded computing devices. The devices can range from cell phones to RFID readers, GPS devices to tablets, embedded control systems in cars to weather stations. In an IoT environment, a door would have the ability to connect with the fire alarm, or your chair would communicate with your home lights, or a car would communicate with the parking space. In the context of this book, we focus on single-board computers (SBCs) as the main IoT hardware components for acquiring and presenting indoor information. This chapter elaborates on different types of SBCs that can be used for acquiring and presenting information regarding building elements, indoor equipment and indoor spaces.

4.1 Introduction

In the future, the Internet will not only be a communication medium for people, it will in fact be a communication environment for devices. Internet of Things (IoT) is defined as a dynamic global network infrastructure with self-configuring capabilities based on standards and interoperable communication protocols. Physical and virtual things in an IoT have identities and attributes and are capable of using intelligent interfaces and being integrated as an information network (Li et al. 2014). The overall concept is known as the IoT. In an IoT environment, a door would have the ability to connect with the fire alarm, or your chair would communicate with your home lights or a car would communicate with the parking space; the list can become longer and longer, and is only limited by your imagination. The present can be regarded as the start of the IoT era. IoT covers the utilization of sensors and near-field communication hardware such as RFID or

© The Author(s) 2015   43
U. Isikdag, *Enhanced Building Information Models*,
SpringerBriefs in Computer Science, DOI 10.1007/978-3-319-21825-0_4

NFC, together with embedded computing devices. The devices can range from cell phones to RFID readers, GPS devices to tablets and embedded control systems in cars to weather stations. In fact, within the context of this book, we will only focus on the single-board computers (SBCs) that facilitate the provision of information acquired from the environment, as (i) SBCs are quite reachable and easy to use and test for development purposes when compared with embedded systems of industrial automation, and (ii) patterns and communication methods explained in this book are based on SBCs as the components of the hardware layer.

SBCs are devices that are developed as proof of concept and experimental tools. These devices have the ability to acquire information from its surroundings by sensors, either embedded in them/or connected to them. SBCs form a solid hardware infrastructure for facilitating the development of the IoT software. The term SBC is used to define computers that consist of a single circuit board memory and the processor. Most SBCs have I/O interfaces where different kinds of sensors can be plugged in. These computers usually do not have extension slots like regular PCs, and the processors used in them are usually of low cost. The size of these devices ranges from a matchbox to a playing card. Some of them have USB and memory card interfaces. SBCs can run versions of embedded Linux and even Windows, while some only have programmable microprocessors which provide output to their proprietary workbench. Recent developments have shown that SBCs can be used as Web servers, or even as a node in cloud clusters. At the time of writing this book, there were more than 20 different types of SBCs produced by different vendors in different parts of the world. In the following sections, we summarize the key efforts in the field of SBCs, which today act as the foundational layer for development of IoT concepts and IoT software layers.

4.2 Arduino Development Boards

One of the most well-known SBC series is the Arduino development boards. Arduino boards were developed to act as development environments or affordable embedded computers which have the ability to acquire information through sensors/and control actuators that are connected to these boards. The Arduino boards include a basic microcontroller and a development framework/workbench for developing the software.

The boards can take inputs from various sensors such as heat, luminance, pressure, magnetic field, proximity and so on and can be used to control various actuators for lights, motors, etc. Thus, the boards are used in education for robotics. They can be obtained from various dealers or can be developed in-house by utilizing the provided (open-hardware) schemas. The common development environment for Arduino is the Processing IDE. As mentioned in the Arduino Web Site (2015), the Arduino programming language is an implementation of Wiring, a similar physical computing platform, which is based on the Processing multimedia programming environment. There are different models of Arduino starting from

Uno to Yun (Figs. 4.1 and 4.2), in different sizes, equipped with different hardware components and designed for different purposes, ranging from robotics to wearable computing. The main advantage that makes Arduino so popular among its competitors is its user-friendly software environment, multinational user support, price and simplicity in developing the platform projects. Arduino IDEs can run cross-platform in multiple operating systems and it is possible to extend the software depending on user requirements. The Arduino hardware development efforts are a part of the open-hardware initiative, so it is possible to build up an Arduino from scratch if you have a certain level of electronics knowledge. Arduino Uno is the most popular model of the Arduino hardware among development communities, while Arduino Yun, which has been released recently, is gaining increasing popularity as it can be used as a micro IoT server, where information gathered from sensors can be directly served over the Web server software which can be installed on the Linux Distribution residing in the Arduino Yun.

There are also many Arduino-compatible boards capable of being programmed using the Arduino programming framework. Another interesting aspect of the Arduino boards is their extensibility. The Arduino boards use the concept of shields, which is similar to the concept of interfaces in object-oriented programing languages; in fact, the interface here is a hardware component instead of a soft-interface in programming languages. A shield can be defined as a hardware component and once plugged to an Arduino board, the board becomes capable of utilizing extra hardware resources. Examples include the Ethernet interface, Wi-Fi interface, a GPRS interface and a GPS interface. Further information about these SBCs can be obtained from Arduino Web Site (2015).

Fig. 4.1 Different models of Arduino (Arduino Web Site 2015). *Source* http://creativecommons.org/licenses/by-sa/3.0/legalcode

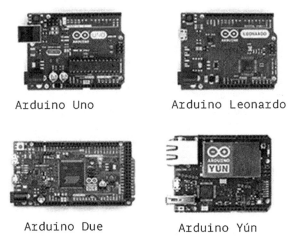

Arduino Uno   Arduino Leonardo

Arduino Due   Arduino Yún

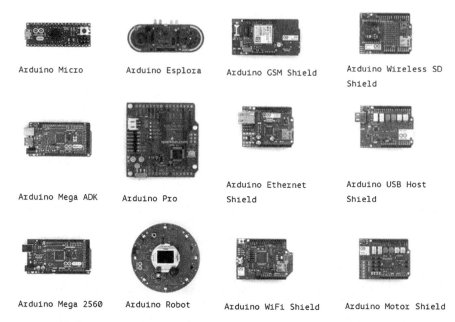

Arduino Micro   Arduino Esplora   Arduino GSM Shield   Arduino Wireless SD
                                                                   Shield

                                            Arduino Ethernet   Arduino USB Host
Arduino Mega ADK   Arduino Pro   Shield   Shield

Arduino Mega 2560   Arduino Robot   Arduino WiFi Shield   Arduino Motor Shield

Fig. 4.2 Different models and shields of Arduino (Arduino Web Site 2015). *Source* http://creativecommons.org/licenses/by-sa/3.0/legalcode

4.3 BeagleBoard

The BeagleBoard SBC is a product of a US-based non-profit foundation which aims to provide and support training in open-source hardware and software. All designs of the board are fully open source and available to the interested parties. The BeagleBoard is known as a low-cost SBC, with fan-less design.

Similar to Android SBC, this board also has pins for acquiring information from sensors and interacting with the actuators. The BeagleBoard is designed to work with Linux distributions and it is also capable of running the Android operating system. The BeagleBoard capes are hardware interfaces similar to Arduino shields, and one can extend the BeagleBoard hardware by using these capes. There are more than 80 capes today and the number is still growing. Examples of the latest releases of the boards and two capes are shown in Fig. 4.3. Further information about these SBCs can be obtained from the BeagleBoard Web Site (2015).

Fig. 4.3 BeagleBoard hardware and capes (BeagleBoard Web Site 2015). *Source* http://creativecommons.org/licenses/by-sa/3.0/legalcode

4.4 CubieBoard

The CubieBoard is an SBC which is produced in China. It is a popular alternative to other SBCs and uses an AllWinner processor, which is a popular processor for low-cost tablets. The board is able to run several versions of Linux, including the Cubian which is specifically developed for the CubieBoard (Fig. 4.4).

The CubieBoard has four different models: CubieBoard1, CubieBoard2, Cubietruck and CubieBoard4. The production team of the CubieBoard had managed to run an Apache Hadoop computer cluster using the Lubuntu Linux distribution. Further information about these SBCs can be obtained from the CubieBoard Web Site (2015).

Fig. 4.4 CubieBoard (Wikimedia Commons 2015). *Source* http://commons.wikimedia.org/wiki/
File:Cubieboard.jpeg

4.5 Raspberry Pi

The Raspberry Pi is the well-known, low-cost SBC that standard PC accessories such as keyboard, monitor, mouse can be plugged into. The Pi efforts started in 2006 by the University of Cambridge staff with the idea of developing a tiny and affordable computer for kids. As indicated in the Raspberry Pi Web Site (2015), by 2008 processors designed for mobile devices became powerful enough to provide excellent multimedia; the idea has become more realizable. Today, Raspberry Pi's capabilities are similar to a PC, as it allows browsing the Internet, playing games, playing HD videos, working with word processing and spreadsheet applications. In addition to PC-like capabilities, the Raspberry Pi is able to acquire information from sensors and interact with the actuators. There are wide ranges of experimental projects where the SBC is used to monitor weather conditions or outside conditions at night with infrared camera input (Fig. 4.5).

The current models of Raspberry Pi are Model A+, Model B+ and Pi 2 Model B. Today, many research projects also use Raspberry Pi as Web servers. The board can be extended with modules, such as the cellular module, which provides the board the capability of communication through cell (mobile phone) networks. The Raspberry Pi runs Raspbian (a free operating system based on Debian Linux and optimized for the Raspberry Pi hardware). Today, there are Raspberry Pi alternatives with similar architecture and capabilities, such as the Orange Pi or the Banana Pi. Further information about these SBCs can be obtained from the Raspberry Pi Web Site (2015).

Fig. 4.5 Raspberry Pi 2 current model (Wikimedia Commons 2015). *Source* http://commons. wikimedia.org/wiki/Raspberry_Pi#/media/File:Raspberry_Pi_2_Model_B_v1.1_top_new.jpg

4.6   Orange Pi

The Orange Pi is an open-source single-board computer alternative to Raspberry Pi. As other Raspberry Pi compatible SBCs, the Orange Pi is capable of running the Raspbian operating system. In addition, the Orange Pi can run Android, Ubuntu and Debian operating systems. Similar to Raspberry Pi, Orange Pi uses the Orange Pi AllWinner processor. The recent models of Orange Pi include Orange Pi Plus, Orange Pi 2 and Orange Pi Mini 2. Further information about these SBCs can be obtained from the Orange Pi Web Site (2015).

4.7   UDOO Board

The UDOO board project was launched as a Kickstarter project and reached its funding goal in 40 h. The UDOO is an SBC with an integrated microcontroller which is Arduino compatible. The main strength of the board comes from running both the Arduino development framework and Linux, and the Android operating systems. Today, the Yun model of the Arduino has similar capabilities. The UDOO board was developed with the aim of supporting (i) the education of computer science and (ii) R&D projects related to IoT. The UDOO board also has the ability to work with Arduino-compatible shields. The Arduino programming framework can be used to program the Arduino-compatible microcontroller in the UDOO board. The UDOO board has the ability to get information from sensors and interact with the actuators in a similar fashion as Arduino. Further information about this SBC can be obtained from the UDOO Web Site (2015).

4.8 Netduino Board

The Netduino board has been developed as a response to the demand from .NET developers following their intention to code for SBCs that are similar to Arduino by using the .NET environment. Although the Netduino board is produced with similarities to Arduino, in terms of microcontroller the SBCs have differences. The main difference in the Netduino hardware is that it can be programmed using the Microsoft .NET development environment. As there are many developers who are familiar with the .NET framework, the Netduino SBCs provides them an easy-to-use interactive environment to start programming for IoT. Another advantage of Netduino is that it is compatible with Arduino shields and can be extended using any of the Arduino/Arduino-compatible shields. Currently, Netduino boards have three models (see Fig. 4.6). Netduino 2 does not have networking capabilities, but Netduino plus 2 and Netduino Go have the ability to connect to the network/Internet. Recent developments have demonstrated that it is

Fig. 4.6 Netduino plus single-board computer (Wikimedia Commons 2015). *Source* http:// commons.wikimedia.org/wiki/Category:Netduino#/media/File:Netduino_Plus.jpg

possible to use Netduino as Web servers to serve web pages or to serve the information acquired from the sensors that are connected to the Netduino. Further information about these SBCs can be obtained from the Netduino Web Site (2015).

4.9 Intel Galileo and Edison

The Intel Galileo is an Arduino-certified development and prototyping board based on the Intel architecture. The SBC is designed for makers, students, educators and DIY electronics enthusiasts. The Intel Galileo Web Site (2015) indicates that the Intel Galileo boards complement and extend Arduino to deliver more advanced computational functionality to those already familiar with Arduino prototyping tools. The Intel Galileo development board is designed to be hardware-, software- and pin-compatible with a wide range of Arduino shields and additionally allows users to incorporate Linux firmware calls in their Arduino sketch programming. Similar to UDOO boards, it is possible to program Galileo using the Arduino development framework. Furthermore, the Intel IoT Developer Kit for Intel Galileo Gen 2 adds C, C++, Python and Node.js/Javascript support for developing connected sensors and IoT applications. SBC is capable of running the Yocto Linux operating system. In addition to the open-source Yocto Linux, Intel Galileo Gen 2 supports VxWorks (RTOS), and now Microsoft Windows is supported directly by Microsoft (Fig. 4.7).

Fig. 4.7 Intel Galileo single (Intel Maker Web Site 2015). Grant of permission by Intel Corporation

Fig. 4.8 Intel Edison (Intel Maker Web Site 2015). Grant of permission by Intel Corporation

Intel Edison is an SBC developed to support prototyping for new product development for the IoT. The main focus of Intel Edison is wearable computing (Fig. 4.8).

Further information about these SBCs can be obtained from the Intel Maker Web Site (2015).

4.10 Radxa Rock

The Radxa Rock is recognized as one of the powerful SBCs in the market capable of doing PC tasks efficiently. As indicated in the Radxa Rock web site, the SBC is shipped with Android and Ubuntu/Linaro and has the dual boot option on the NAND flash (onboard storage). All PC accessories can work with this SBC. Rock chips are also used in SBCs that are used to watch TV channels (Fig. 4.9).

Fig. 4.9 An android TV stick with rockchip (Wikimedia Commons 2015). *Source* http://commons.wikimedia.org/wiki/File:MK809III_V1.0_130606_inside_front.jpg?uselang=tr

The IoT software development research in the academic world today is much facilitated by the existence and utilization of SBCs. These devices form well-defined interfaces between the sensors and other software layers such as middle-tier data acquisition components or integration portals. While these SBCs form the hardware-backbone of the IoT, the integration efforts from the software side are today dominated by integration portals and platforms. In the following chapter of this book, we will focus on the software-backbone components of the IoT, and specifically on the integration platforms. Further information about these SBCs can be obtained from the Radxa Rock Web Site (2015).

References

Arduino Web Site (2015). Available online at: http://www.arduino.cc
Beagle Board Web Site (2015). Available online at: http://www.beagleboard.org
Cubie Board Web Site (2015). Available online at: http://cubieboard.org
Intel Maker Web Site (2015). Available online at: http://www.intel.com/content/www/us/en/do-it-yourself/maker.html
Li, S., Xu, S.D., Zhao, S.: The internet of things: a survey. Inf. Syst. Front. **17**(2), 243–259 (2014)
Netduino Web Site (2015). Available online at: http://www.netduino.com/
Orange Pi Web Site (2015). Available online at : http://www.orangepi.org
Radxa Rock Web Site (2015). Available online at: http://www.radxa.com
Raspberry Pi Web Site (2015). Available online at: https://www.raspberrypi.org
UDOO Web Site (2015). Available online at: http://www.udoo.org/
Wikimedia Commons (2015). Available online at http://commons.wikimedia.org/wiki/Main_Page

Chapter 5
Internet of Things: Software Platforms

Abstract The Internet of Things (IoT) architectures do not only consist of hardware. The IoT hardware would require operating systems to work and also need to implement communication protocols to communicate with other devices and humans. Furthermore, there are middleware components that facilitate communication and exchange of information between devices. In IoT architectures, integration portals play an important role in combining and integrating information acquired from multiple devices and presenting this information to the users. This chapter provides detailed information on the software side of IoT.

5.1 Introduction

The software side of the Internet of Things (IoT) is more complicated than the hardware side. The hardware components include embedded systems, sensors and SBCs. As mentioned in the previous chapter, IoT covers the utilization of sensors and near-field communication hardware such as RFID or NFC, together with embedded computing devices. In fact, in the previous chapter we only focused on single-board computers as they are easy to reach and operate, and as patterns and communication methods explained in this book focus on the use of these devices. In this chapter, the software side of the IoT approach is summarized from a broader perspective.

As depicted in Fig. 5.1 the software component of the IoT is formed by different layers. The lowest software layer is the operating system(s) of the hardware components. Either an embedded system or an SBC hardware would mostly be controlled by an operating system. There are exceptions to this where the hardware directly communicates with the middleware or a programming framework. The protocols play a role in enabling communication between different software layers. The middleware and development frameworks form the core layer of the IoT

© The Author(s) 2015
U. Isikdag, *Enhanced Building Information Models*,
SpringerBriefs in Computer Science, DOI 10.1007/978-3-319-21825-0_5

Fig. 5.1 IoT software
components

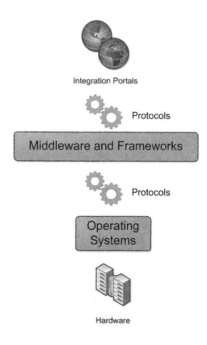

Integration Portals

Protocols

Middleware and Frameworks

Protocols

Operating
Systems

Hardware

software. These components act as a bridge between devices to enable M2M communication, to facilitate provision of the information acquired from the surrounding environment to the users or to the integration portals. The integration portals are at the top of the other layers; these portals are actively used in integrating information from multiple devices (hardware or soft sensors) and presenting this information to the user. In the following sections we focus on key examples of the software components in each layer.

5.2 Operating Systems

5.2.1 Mobile Operating Systems

Most well-known embedded operating systems run on the mobile phones of today. There are three main operating systems in the field, Android, iOS and Windows Phone. Android is a Linux-based operating system developed by Google. The OS's focus is enabling user interaction in touch screen devices. Apart from the mobile phone, the Android OS is also used to control tablet computers, PCs, TV sets, digital cameras, game consoles and cars. The source code of the Android OS is open. The iOS is the mobile operating system for Apple devices. The iOS is used to interact with Apple's mobile phones and tablet computers. The Windows Phone uses Windows operating systems for managing mobile phones. Starting from

Windows 8 the tablet computers are able to run the Windows operating system which is designed for multiple device types (such as tablets, PCs and so on.)

5.2.2   OpenWRT

OpenWRT is the operating system used with the most advanced model of the Arduino, i.e. the Arduino Yun. The OpenWRT Web Site (2015) defines the OS as a highly extensible GNU/Linux distribution for embedded devices. The main focus of the OS is wireless routers. OpenWRT is free to use and open source and the system is developed as a result of a community-driven effort. Apart from the wireless routers that the OS is developed for and the Arduino Yun SBC, a 3D printing project (Doodle3D) uses OpenWRT as the OS of their environment.

5.2.3   Windows Embedded

The Windows Embedded operating system is developed by Microsoft to control and manage the embedded (industrial) devices. The graphical user interface of the latest version of Windows Embedded is similar to the latest version of Windows. The system can be used to control POS systems/devices, medical devices in healthcare systems and industrial computers used in manufacturing process. Furthermore, some car manufacturers use this OS at indoor entertainment systems.

5.2.4   Raspbian

The operating system Raspbian is tightly coupled with Pi compatible hardware. As indicated by the Raspbian Web Site (2015) it is a free operating system based on Debian optimized for the Raspberry Pi hardware. The operating system is a set of basic programs and utilities that make your Raspberry Pi run. Raspbian comes with over 35,000 packages, pre-compiled software bundled in a nice format for easy installation on the Raspberry Pi. The Raspberry Pi compatible hardware such as Orange Pi can also run the Raspbian OS. There also exists a server edition of the Raspbian OS.

5.2.5   Contiki OS

Contiki OS has been developed with the focus on integration of wireless low-power/low-memory IoT nodes (Contiki Web Site 2015). These hardware components can be networked together to form a wireless sensor network. City applications such as street lighting systems, sound level monitoring, pollution monitoring systems make use of this OS. The Contiki OS supports IPv4 and IPv6 networking. A popular tool of the Contiki system is its network simulator Cooja. The network simulator Cooja runs on Linux operating system, and is able to simulate the network nodes.

5.2.6 RIOT OS

Baccelli et al. (2012) identified RIOT OS as a microkernel-based operating system matching the various software requirements for IoT devices. RIOT OS has an adaptive network stack, providing full-fledged IPv6 as well as protocols targeting more constrained networks such as 6LoWPAN or RPL. It is indicated that by providing the same developer-friendly API across all platforms (from 16-bit micro-controllers to 32-bits processors) and by simultaneously providing key features such as real-time capabilities and energy efficiency, RIOT OS is able to power a wide spectrum of IoT devices.

5.2.7 Tiny OS

The Tiny OS is a BSD clone that is designed for low-power wireless devices which form (i) the backbone of the wireless sensor networks (ii) and when connected to the internet through M2M communication form the backbone of an IoT architecture. These low-power devices are used in smart-city applications, smart buildings as smart devices. The TinyOS supports IPv6 stack, 6LoWPAN or RPL protocols as well. A global community supports the development of the TinyOS (TinyOS Web Site 2015).

5.2.8 Free RTOS

The RTOS acronym stands for Real-Time Operating System. As explained in the Free RTOS Web Site (2015) "...*most operating systems appear to allow multiple programs to execute at the same time. This is called multi-tasking. In reality, each processor core can only be running a single thread of execution at any given point in time. A part of the operating system called the scheduler is responsible for deciding which program to run when, and provides the illusion of simultaneous execution by rapidly switching between each program. The scheduler in a Real-Time Operating System (RTOS) is designed to provide a predictable (deterministic) execution pattern. This is particularly of interest to embedded systems as embedded systems often have real-time requirements. A real-time requirements is one that specifies that the embedded system must respond to a certain event within a strictly defined time (the deadline). A guarantee to meet real-time requirements can only be made if the behaviour of the operating system's scheduler can be predicted (and is therefore deterministic)."* Free RTOS is a component of RTOS that is designed to be small enough to run on a single micro-controller.

5.3   Hardware and Software Bundles

There are also some specific bundles where the hardware components require specific software or frameworks to interact with the hardware nodes. This section will present two examples of these bundles.

5.3.1   Spark.IO

The Spark.IO is a set of hardware and software components including micro-controllers, connectivity modules (i.e. modules for Wi-Fi and Cell Networks) and set of extension components (i.e. the shields). The software tools include a Wiring framework to develop code for the hardware components (coding is very similar to Arduino development environment), a mobile app for remote control and monitoring, and a Web-based IDE. The software components also provide support for the use Node.JS framework for interacting with the Spark nodes (Spark.IO Web Site 2015).

5.3.2   Open Mote

The open hardware for the IoT project, Open Mote, provides a set of core modules and an interface board which can interface various sensors and operate autonomously as a node. The hardware utilizes the XBee standard to connect to the other nodes. As stated by the Open Mote Web Site (2015) the Open Mote hardware has been designed to support various open-source software stacks specifically designed for the IoT, that is, Contiki and OpenWSN. In addition, the Open Mote hardware is also known to run FreeRTOS, a real-time operating system for embedded devices and RiOT, another fully open-source RTOS designed for the IoT (OpenMote Web Site 2015).

5.4   Messaging Standards and Protocols

The TCP/IP protocol suite offers many possibilities for the enablement of IoT, but in fact, in several different layers of the IoT network architecture, there are protocols that specifically focus on enabling and facilitating the communication of nodes in an IoT network. As presented in Fig. 5.2 protocols such as RPL and 6LowPAN at the network layer, COAP at the application layer are specifically defined for facilitating M2M communication. There are also protocols that benefit from TCP/IP application and transport layers for enabling message exchange between devices such as MQTT and XMPP. In this section, we elaborate on IoT-specific communication protocols in the upper layers of the network architecture that support enablement and utilization of IoT. It should be noted that (lower) physical layer protocols such as IEEE 802.15.4, Zigbee, Bluetooth and NFC will not be elaborated in this section as they are not within the context of this book.

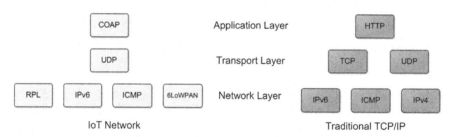

Fig. 5.2 Communication protocols of IoT network versus traditional TCP/IP

5.4.1   RPL Protocol

The RPL protocol acronym stands for low-power and Lossy Network (LLN) routing protocol. According to the definition in the standard (RPL IETF 2015) LLNs consist largely of constrained nodes with limited processing power, memory and sometimes energy. It is assumed that in Lossy Networks a vast number of nodes are interconnected with unstable links supporting only low data rates. The traffic patterns in Lossy Networks are point to multipoint or vice versa. The RPL protocol is an IPv6 routing protocol that supports routing in Lossy Networks.

5.4.2   6LoWPAN Protocol

6LoWPAN is the acronym for IPv6 over low-power wireless personal area networks. The protocol was originated with the idea of facilitating routing between low-power devices with limited processing capacities in the IoT environment. The idea behind this routing protocol was related with connecting devices with wireless connectivity and low data rates. The idea applies to many nodes that can broadcast and share information during a production process or within the smart-city grid.

5.4.3 CoAP Protocol

Constrained Application Protocol (CoAP) is an application layer protocol. The focus of the protocol is resource-constrained Internet devices, such as WSN nodes. The CoAP protocol follows a RESTful-style architecture, and provides similar interaction approaches to HTTP methods (such as GET, PUT, POST, DELETE...). As mentioned in the CoAP Web Site (2015) the focus of the protocol is on facilitating M2M interactions such as smart energy and smart homes. The CoAP queries are similar to RESTful queries, i.e. it is a representation (of a Web resource) that the query targets, and this resource can provide another representation or get updated or deleted as a result of a CoAP query. A sample query syntax is "GET coap:// 192.18.1.1:22".

5.4.4 MQTT Protocol

The acronym MQTT stands for Message Queue Telemetry Transport. The MQTT is a lightweight messaging protocol that is used over the TCP/IP protocol stack. Similar to protocols explained in this section, this protocol is designed for low-power networks with limited resources, and for networks that are assumed to be less reliable. MQTT provides a publish/subscribe model and, as any publish/subscribe protocol it implements a message broker. As the protocol is based on top of TCP/IP, both client and broker need to have a TCP/IP stack. The aim of the protocol is minimization of bandwidth use, but assuring some degree of delivery. The protocol can be used in M2M communication, specifically when power consumption and bandwidth is a key issue. TCP/IP Port 1883 is used for MQTT protocol communication.

5.4.5 XMPP Protocol

The protocol formerly named Jabber is an XML-based protocol for message-oriented middleware. The extensible messaging and presence protocol was originally developed for message exchange for instant messaging applications. XMPP provides a client–server model instead of a peer-to-peer approach. In fact, there is no centrality of servers as it is usual for intended parties to run their own Jabber servers. The publish/subscribe model provided by the XMPP protocol is beneficial for IoT architectures. XMPP extensions are used in facilitating communication in IoT systems. Examples include the use of XMPP in automatic power metre reading and in energy efficiency research.

5.5 Middleware and Frameworks

The IoT middleware and frameworks form the core layer of the IoT software components. The IoT middleware accomplishes the goal of enabling machine-to-machine and machine-to-user/user-to-machine communication and interaction. The middleware layer mainly focuses on orchestrating the exchange of information and interactions between (i) devices, (ii) between devices and other software layers. The IoT frameworks can have many functions ranging from providing development/coding interfaces to hardware, to providing component containers to interact with real or simulated (virtual) sensors. This section will focus on some key examples of IoT middleware and frameworks.

5.5.1 AllSeen Alliance and AllJoyn

AllSeen Alliance is the broadest, cross-industry open-source effort to advance the IoT and Internet of Everything beyond the connected home. The initiative includes companies including consumer electronics manufacturers, home appliance makers, automotive companies, IoT cloud providers, enterprise technology companies, innovative startups, chipset manufacturers, service providers, retailers and software developers. The AllSeen Alliance manages the AllJoyn open-source project (AllSeen Alliance Web Site 2015). According to the PC World (2015) AllJoyn is an open-source software framework for IoT devices to discover and communicate with each other. It is designed to allow for a single app or service to interact with multiple IoT devices, such as lights, heating systems, security cameras and wearables.

5.5.2 Eclipse IOT Frameworks and Services

Eclipse IoT provides a set of frameworks and services which can be regarded as building blocks that sit on top of open standards and protocols (Eclipse IOT Web Site 2015). Two well-known frameworks of Eclipse are Kura and Mihini. Kura is a Java-based framework for IoT gateways. Kura APIs offer access to the underlying hardware (serial ports, GPS, watchdog, GPIOs, I2C, etc.), management of network configurations, communication with M2M/IoT integration platforms and gateway management (Kura Web Site 2015). On the other hand, the Mihini project delivers an embedded run time running on top of Linux that exposes a high-level API for building machine-to-machine applications (Mihini Web Site 2015). Eclipse offers two IoT service components, Smart Home and SCADA. The Smart Home components focus on smart home and ambient-assisted living (AAL) solutions. The project focuses on services and APIs for data handling, rule engines, declarative user interfaces and persistence management. SCADA is a set of tool development libraries, interface applications, mass configuration tools, front-end and back-end applications to connect different industrial devices.

5.5.3   IoTSyS Middleware

IoTSyS provides a middleware layer to enable the integration of objects in the IoT architecture. It is focused on facilitating communication for embedded devices addressed using IPv6. The focus of the middleware is on enhancing the interoperability of smart objects. The middleware utilizes the 6LoWPAN and COAP and XML interchange over Web services to interact with sensors and actuators. As indicated in the IoTSyS Web Site (2015), the IoTSyS middleware aims at providing a gateway concept for existing sensor and actuator systems found nowadays in home and building automation systems, a stack which can be deployed directly on embedded 6LoWPAN devices and further addresses security, discovery and scalability issues.

5.5.4   IoTivity Framework

The IoTivity framework aims to create extensible and interoperable multi-device multi-platform software architecture for integration of smart devices. The APIs of the framework use a resource-oriented RESTful architecture. The framework supports information exchange based on COAP model, has a publish-subscribe model and provides a device discovery mechanism. Resource-oriented architecture supports GET and PUT requests; other requests such as OBSERVE are possible. The framework supports input from the soft sensors. IoTivity provides framework and services to implement a controller with smart home data model (IoTivity Web Site 2015).

5.5.5   OpenIoT Project

The project name stands for open-source cloud solution for IoT. The aim of the project is to provide an open-source middleware for acquiring information from the sensor clouds regardless of the type of sensors used. As stated in the OpenIoT Web

Site (2015), OpenIoT focuses on the development of middleware for sensors and sensor networks, ontologies, semantic models and annotations for representing Internet-connected objects, along with semantic open-linked data techniques, cloud/utility computing, including utility-based security and privacy schemes. OpenIoT middleware infrastructure allows flexible configuration and deployment of algorithms for collection, and filtering information streams stemming from Internet-connected objects.

5.5.6   Macchina.IO

The Macchina.IO is a toolkit for developing IoT applications for SBCs that run Linux OS such as Arduino Yun, Raspberry Pi or so on. The framework provides modular JavaScript and C++ run time environment which enables to acquire information from multiple sensors and present it to cloud services. The framework utilizes RESTful architecture and the MQTT protocol for communication.

5.6   Integration Portals

The integration portals can be defined as web sites that contain interfaces (i) to facilitate M2M integration, (ii) to enable visualization of information obtained from sensors, (iii) to facilitate user interaction with actuators, (iv) to provide a service based on information acquired from sensors, (v) to update web resources based on information acquired from the sensors. Thus, these sites aim to facilitate multiple dimensions of IoT integration and can be termed as Integration Portals.

5.6.1   Xively

The portal formerly called Pachube and later COSM, is a platform as a service for IoT. It is one of the pioneer Web integration platforms that provided APIs for different hardware components and SBCs, for enabling M2M and user interaction. Furthermore, the APIs are also useful in enabling discovery of devices and integration of information published by these devices.

As illustrated in Fig. 5.3 Xively has become a commercial platform that provides directory services, data services and business services focused on IoT. The message bus layer deals with the real-time routing and management of messages. Xively API is the core component of the Xively tools and forms a gateway between front-end and back-end software components, mobile applications and IoT nodes (i.e. connected objects). Xively API supports reading and writing data via three resources: feeds, datastreams and datapoints (Fig. 5.4).

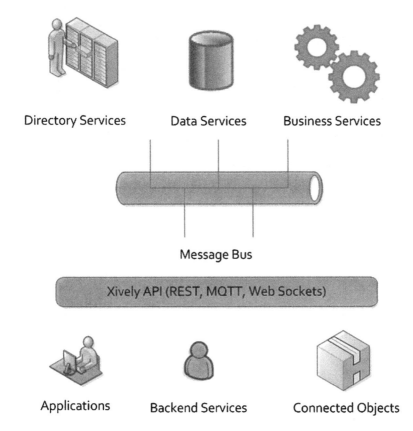

Fig. 5.3 Xively API and services

As mentioned by the Xively Web Site (2015) a feed is the collection of channels (datastreams). A feed's metadata can optionally specify location, tags, whether it is physical or virtual, fixed or mobile, indoor or outdoor, etc. Every device has exactly one feed. A datastream is a bidirectional communication channel that allows for the exchange of data between the Xively platform and authorized devices, applications and services. Each datastream represents a specific attribute, unit or type of information (a variable). Datapoint: A datapoint represents a single value of a datastream at a specific point in time. It is simply a key-value pair consisting of a timestamp and the value at that time. For example requesting the datastream as a PNG image with the following HTTP GET request will generate a customizable graph of the datastream's history as a .png file (Fig. 5.5).

[GET https://api.xively.com/v2/feeds/FEED_ID_HERE/datastreams/DATASTREAM_ID.png]

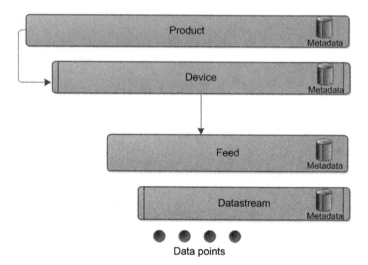

Fig. 5.4 Xively feeds and datastreams

Fig. 5.5 An illustration of the graph of a datastream generated as a result of a GET request

```
01.  {
02.      "id":"example",
03.      "current_value":"500",
04.      "at":"2013-05-06T00:30:45.694188Z",
05.      "max_value":"500.0",
06.      "min_value":"333.0",
07.      "version":"1.0.0"
08.  }
```

Fig. 5.6 JSON output of a datastream generated as a result of GET request

```
01.    {
02.        "version":"1.0.0",
03.          "datastreams" : [ {
04.               "id" : "example",
05.               "datapoints":[
06.               {"at":"2013-04-22T00:35:43Z","value":"42"},
07.               {"at":"2013-04-22T00:55:43Z","value":"84"},
08.               {"at":"2013-04-22T01:15:43Z","value":"41"},
09.               {"at":"2013-04-22T01:35:43Z","value":"83"}
10.               ],
11.               "current_value" : "40"
12.          }
13.      ]
14.    }
```

Fig. 5.7 JSON input for a PUT request

Similar request/response for the JSON output of a datastream would be as below (Fig. 5.6).

[GET https://api.xively.com/v2/feeds/FEED_ID_HERE/datastreams/DATASTREAM_ID.json]

A JSON request for writing multiple datapoints to singe datastream would be as below (Fig. 5.7).

[PUT https://api.xively.com/v2/feeds/FEED_ID_HERE/]

5.6.2 Paraimpu

The Paraimpu describes itself as a social tool defined to connect things with the aim of creating personal applications for IoT. The portal focuses on three specific directions. As mentioned in the Paraimpu Web Site (2015) these are (a) Connect your things, such as sensors, motors, micro-controllers like Arduino, home appliances, lighting systems or whatever you want, talk with the Web. (b) Compose and inter-connect them together or to social networks to interact with them or publish their feeds in social networks. (c) Share and let other people use produced data in their own connections, enabling real, social, physical-virtual web mashups. The Paraimpu workspace (illustrated in Fig. 5.8) is used to connect the information gathered from sensors or SBCs (such as Arduino) to real actuators in SBCs or virtual actuators such as Twitter post publisher. It is also possible to define M2M communication services through the Paraimpu workspace.

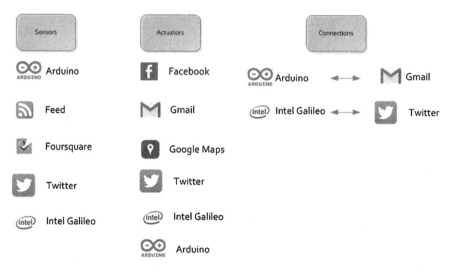

Fig. 5.8 Paraimpu workspace illustration

5.6.3   Dweet.IO

The Dweet.IO is a simple publish-subscribe service for things in an IoT environment (Dweet.IO Web Site 2015). The service is very simple to use, i.e. only requires a GET request to interact with the API. A Thing (for instance a SBC) can issue the GET request for a Hello World with the following query parameters.

```
01.   {
02.       "this": "succeeded",
03.       "by": "getting",
04.       "the": "dweets",
05.       "with": [
06.           {
07.               "thing": "my-thing-name",
08.               "created": "2014-01-15T18:41:17.166Z",
09.               "content": {
10.                   "this": "is cool!"
11.               }
12.           },
13.           {
14.               "thing": "my-thing-name",
15.               "created": "2014-01-15T18:41:01.583Z",
16.               "content": {
17.                   "hello": "world",
18.                   "foo": "bar"
19.               }
20.           }
21.       ]
22.   }
```

Fig. 5.9 Dweet.IO JSON response to GET request

[GET https://dweet.io/dweet/for/my-thing-name?hello=world&foo=bar]

In order to get a dweet (i.e. device tweet) the following GET request would be sufficient.

[GET https://dweet.io/get/latest/dweet/for/my-thing-name]

The JSON response would be as shown in Fig. 5.9.

5.6.4   Freeboard.IO

The Freeboard.IO integration platform provides easy-to-build visualization for information acquired from IoT components (Freeboard.IO Web Site 2015). This includes the generation of dashboards by making use of the drag-and-drop interface. The Freeboard API can be seamlessly integrated for Dweet.IO output and can generate visualization from the dweets. Figure 5.10 depicts an illustration of a dashboard designed for monitoring indoor conditions of a residential property.

Fig. 5.10 Freeboard.IO board user interface illustration—i.e. presenting information acquired from multiple indoor sensors

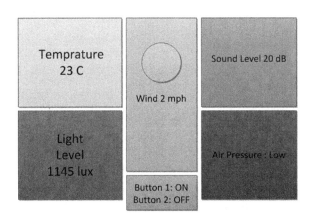

References

AllSeen Alliance Web Site (2015). Available online at: https://allseenalliance.org
Baccelli, E., Hahm, O., Wahlisch, M., Gunes, M., Schmidt, T.: RIOT: One OS to Rule Them All in the IoT. (Research Report) RR-8176, (2012). Available online at: https://hal.inria.fr/hal-00768685v3
CoAP Web Site (2015). Available online at: http://coap.technology
Contiki Web Site (2015). Available online at: http://www.contiki-os.org
Dweet.IO Web Site (2015). Available online at: http://dweet.io
Eclipse IOT Website (2015). Available online at: http://iot.eclipse.org
Free RTOS Web Site (2015). Available online at: http://www.freertos.org
Freeboard.IO Web Site (2015). Available online at: http://freeboard.io

IoTivity Web Site (2015). Available online at: https://www.iotivity.org
IoTSyS Web Site (2015). Available online at: https://code.google.com/p/iotsys
Kura Web Site (2015). Available online at: https://eclipse.org/kura
Mihini Web Site (2015). Available online at: https://eclipse.org/mihini
OpenIoT Website (2015). Available online at: http://openiot.eu
OpenMote Web Site (2015). Available online at: http://www.openmote.com
OpenWRT Web Site (2015). Available online at: https://openwrt.org
Paraimpu Web Site (2015). Available online at: https://www.paraimpu.com
PC World (2015). Available online at: http://www.pcworld.com/article/2866252/alljoyn-iot-platform-reaches-out-to-the-internet-for-remote-control.html
Raspbian Web Site (2015). Available online at: http://www.raspbian.org
RPL IETF (2015). Available online at: https://tools.ietf.org/html/rfc6550#page-8
Spark.IO Web Site (2015). Available online at: http://www.spark.io
TinyOS Web Site (2015). Available online at: http://www.tinyos.net
Xively Web Site (2015). Available online at: http://xively.com

Chapter 6
Advanced SOA Patterns for Building Information Models

Abstract Two styles of Web services exist today: Simple Object Access Protocol (SOAP) and REST. Representational State Transfer (REST) is often preferred over the more heavyweight SOAP because REST does not leverage as much bandwidth. REST's decoupled architecture makes it a popular building style for cloud-based APIs, such as those provided by Amazon, Microsoft and Google. This chapter starts with providing technical information about RESTful Web services. Following this, RESTful design patterns for facilitating BIM-based software and Web service architectures are presented.

6.1 Introduction

As indicated in Isikdag and Underwood (2009) Web services can be defined as components and resources that can either be invoked over the Web or reached by standard Web protocols using messages. Exposing a Web service mostly involves enabling the older software (or the data layer of the legacy system) to receive and respond to web message requests for its functionality (Pulier and Taylor 2006). He (2003) indicated that two constraints exist for implementing the Web services: (i) interfaces must be based on Internet protocols such as HTTP, FTP and SMTP and (ii) except for binary data attachment, messages must be in XML. Two definitive characteristics of Web services are mentioned as loose coupling and network transparency (Pulier and Taylor 2006). As explained by the authors, in a traditional distributed environment computers are *tightly coupled*, i.e. each computer connects with others in the distributed environment through a combination of proprietary interfaces and network protocols. Web services in contrast, are *loosely coupled*, i.e. when a piece of software has been exposed as a Web service, it is relatively simple to move it to another computer. On the other hand, as Web services' consumers and providers send messages to each other using open Internet protocols, Web services offer total *network transparency* to those that employ them (Pulier and Taylor 2006). Network transparency refers to a Web service's capacity

© The Author(s) 2015   71
U. Isikdag, *Enhanced Building Information Models*,
SpringerBriefs in Computer Science, DOI 10.1007/978-3-319-21825-0_6

to be active anywhere on a network or group of networks without having any impact on its ability to function. As each Web service has its own Universal Resource Indicator (URI), Web services have similar flexibility to web sites on the Internet. Two styles of Web services exist today: SOAP and REST. Simple Object Access Protocol (SOAP) is a protocol for exchanging XML formatted messages between the networks using Hypertext Transfer Protocol (HTTP). The SOAP protocol has two constraints: (i) except for binary data attachment, messages must be carried by the SOAP protocol and formatted by its rules and (ii) the description (exposed interface) of the service must be defined in the Web Services Description Language (WSDL). The terms REST and RESTful Web services have been coined after the PhD dissertation of Fielding (2000). As explained by Pautasso et al. (2008), REST was originally introduced as an architectural style for building large-scale distributed hypermedia systems. According to REST style, a Web service can be built upon resources (i.e. anything that is available digitally over the Web), their names (identified by uniform indicators, i.e. URIs) representations (i.e. metadata/data on the current state of the resource) and links between the representations. As mentioned by Techtarget (2015), REST is often preferred over the more heavyweight SOAP style because REST does not leverage as much bandwidth. REST is often used in mobile applications, social networking web sites, mashup tools and automated business processes. REST's decoupled architecture makes it a popular building style for cloud-based APIs, such as those provided by Amazon, Microsoft and Google. When Web services use the REST architecture, they are called RESTful Application Programming Interfaces (APIs) or REST APIs. REST decouples information consumers from producers and supports stateless existence where the server does not store the number or state of the clients consuming a Web service. The number of operations is limited with the methods provided by the HTTP protocol. GET, PUT, POST and DELETE are the most commonly used methods (i.e. the number of verbs to be used is limited). In fact, the hierarchical naming convention on REST APIs provides flexibility where each RESOURCE is identified by a universal resource identifier (URI) which is defined by nouns. REST also has downsides. In the world of REST, as mentioned by Techtarget (2015), there is no direct support for generating a client from server-side-generated metadata. SOAP is able to support this with WSDL. In fact, RESTful APIs have the advantage of exchanging plain messages, while SOAP uses XML envelopes, which make the message size larger and message transfer less efficient. Most of the Web services currently implemented have transformed from interfaces of the legacy systems.

6.2 REST in a Nutshell

In order to explain the RESTful architectures it would be good to start from a simple web page request. Once a request is made from a web browser by the user to view a page (such as www.google.com) a GET request is issued according to the

HTTP protocol. A GET request in the HTTP protocol is a request where you can pass a URI including some parameters. For instance,

http://www.somewebsite.org/showbuildingpart.php?building_id=1254&floor_id=12

is a valid GET request with parameters. In fact, the HTTP protocol offers just more than the GET request. Another commonly used request in the web browsers is the POST request. The POST request sends information to a URI but as opposed to the GET request does not include parameters as a part of the request. In a POST request, information that is required by the receiving end is sent within the HEAD or BODY of the message. In fact, the HTTP protocol capabilities are not limited to the GET and POST, but the HTTP protocol is able to send a PUT request to update the REPRESENTATION of a web RESOURCE and DELETE request to delete a resource. There are also other HTTP methods such as HEAD, TRACE, PATCH, CONNECT or OPTIONS, but these will not be elaborated here. The REST architectural principles indicate that the information on the Web can be managed using the HTTP methods, in a similar way that one can manage the Create/Read/Update/Delete (CRUD) operations for a data resource. In a RESTful architecture HTTP methods GET, POST, PUT and DELETE are used to make CRUD-like operations over the web RESOURCES. According to the REST architectural style a Web service can be built by utilizing …

- RESOURCES (i.e. anything that is available digitally over the Web),
- IDENTIFIERS (i.e. URIs),
- REPRESENTATIONS (i.e. current state of the resources).

In a RESTful architecture RESOURCES, REPRESENTATIONS and IDENTIFIERS can be described as below:

- RESOURCE → is a logical object identified by an IDENTIFIER.
- IDENTIFIER → A globally unique ID that points to the RESOURCE.
- REPRESENTATION: Physical source of information that is pointed by the IDENTIFIER.
- A RESOURCE can have multiple representations but a single IDENTIFIER which can only point to a single REPRESENTATION of a RESOURCE at a single point in time.

Once a RESOURCE is created in a RESTful architecture it is constant until it has got DELETED. The variability is on the REPRESENTATION, as the REPRESENTATION can be UPDATED using the HTTP protocol methods. The acronym REST stands for Representational State Transfer. In a RESTful architecture a REPRESENTATION existent in one STATE of a RESOURCE is TRANSFERRED to a web client and this causes the change in the STATE of the web client. It is vital to mention here that the SERVER is STATELESS in RESTful architectures, which means that the CLIENT STATE is not known or maintained by the SERVER, which provides great efficiency in RESTful architectures. Table 6.1 summarizes valid server-side RESOURCE REPRESENTATIONS by examples.

Table 6.1 Resource representations (server side)

Resource	Identifier	Representation	Time	Resource state
Building footprint	http://www.xyz.com/building/ 0001/footprint	footprint.jpg	$t = 0$	$s = i$
Building footprint	http://www.xyz.com/building/ 0001/footprint	footprint.csv	$t = 1$	$s = ii$
Building footprint	http://www.xyz.com/building/ 0001/footprint	footprint.dwg	$t = 2$	$s = iii$
Building facade	http://www.xyz.com/building/ 0001/facade	façade.jpg	$t = 3$	$s = i$
Building facade	http://www.xyz.com/building/ 0001/facade	façade.dwg	$t = 4$	$s = ii$
(Deleted) building facade			$t = 5$	$s = $ null

An example sequence of client interaction in a RESTful architecture is provided in Table 6.2. As indicated in Table 6.2.

1. client states ($s = 1, 3, 5, 7, 9, 11, 13, 15, 17, 19$) are REQUEST states where the client makes an HTTP REQUEST to a RESTful Service,

   (a) to GET REPRESENTATION of a RESOURCE by an IDENTIFIER (i.e. URI)

      i. using HTTP GET ($s = 1, 5, 9, 13, 17$)

   (b) to UPDATE REPRESENTATION of a RESOURCE by an IDENTIFIER

      ii. using HTTP PUT ($s = 3, 7, 15$)

   (c) to CREATE a new RESOURCE by POSTING a REPRESENTATION of it with an IDENTIFIER

      iii. using HTTP POST ($s = 11$)

   (d) to DELETE a RESOURCE by an IDENTIFIER

      iv. using HTTP DELETE ($s = 19$)

2. client states ($s = 2, 6, 10, 14, 18$) are the VIEW states where the information transferred from the server is shown by the client.
3. client states ($s = 4, 8, 12, 16, 20$) are the ACKNOWLEDGEMENT states where the ACKNOWLEDGEMENT on method success or failure sent by the server are shown by the client.
4. in the HTTP REQUEST states the client view stays on the previous state until the request is fulfilled (e.g. in web modern browsers this is the reason that we continue to see a web page until the next page is loaded).

Table 6.2 Client-side interactions within a RESTful architecture

Client state	Client HTTP REQUEST	Client view	Data sent by client
$s = 1$	GET http://www.xyz.com/building/0001/footprint	Nothing	Nothing
$s = 2$	Nothing	footprint.jpg	Nothing
$s = 3$	PUT http://www.xyz.com/building/0001/footprint	footprint.jpg	footprint.csv
$s = 4$	Nothing	HTTP 200	Nothing
$s = 5$	GET http://www.xyz.com/building/0001/footprint	HTTP 200	Nothing
$s = 6$	Nothing	footprint.csv	Nothing
$s = 7$	PUT http://www.xyz.com/building/0001/footprint	footprint.csv	footprint. dwg
$s = 8$	Nothing	HTTP 200	Nothing
$s = 9$	GET http://www.xyz.com/building/0001/footprint	HTTP 200	Nothing
$s = 10$	Nothing	footprint. dwg	Nothing
$s = 11$	POST http://www.xyz.com/building/0001/facade	footprint. dwg	façade.jpg
$s = 12$	Nothing	HTTP 201	Nothing
$s = 13$	GET http://www.xyz.com/building/0001/facade	HTTP 201	Nothing
$s = 14$	Nothing	façade.jpg	Nothing
$s = 15$	PUT http://www.xyz.com/building/0001/facade	façade.jpg	façade.dwg
$s = 16$	Nothing	HTTP 200	Nothing
$s = 17$	GET http://www.xyz.com/building/0001/facade	HTTP 200	Nothing
$s = 18$	Nothing	façade.dwg	Nothing
$s = 19$	DELETE http://www.xyz.com/building/0001/ facade	façade.dwg	Nothing
$s = 20$	Nothing	HTTP 200/ HTTP 204	Nothing

5. An ACKNOWLEDGEMENT of a successful GET request is usually not shown to the users, instead the client state is transferred to the VIEW state where users see the REPRESENTATION of the RESOURCE.
6. Regular HTTP clients such as web browsers do not show HTTP ACKNOWLEDGEMENTS, a RESTful service client is required to observe all states provided in Table 6.2.

As it can be noticed from Tables 6.1 and 6.2 the RESTful requests utilize a hierarchical URI structure. As explained by the RESTful API Tutorial (2015), in addition to utilizing the HTTP verbs appropriately, resource naming is arguably the most debated and the most important concept to grasp when creating an understandable, easily leveraged Web service API. When resources are named well, an API is intuitive and easy to use. Generally a RESTful API is a collection of URIs. A RESTful URI should refer to a resource which is a thing instead of referring to an action. URIs should follow a predictable, hierarchical structure to enhance usability;

hierarchical in the sense that data has structure relationships. This is not a REST rule or constraint, but it enhances the API. If we go back to our first example, URI called with an HTTP GET request

http://www.somewebsite.org/showbuildingpart.php?building_id=1254&floor_id=12

has its counterpart URI in a RESTful API which implies a hierarchical naming convention would be one of the following:

http://www.somewebsite.org/building_id/1254/floor_id/12
http://www.somewebsite.org/building/1254/12

REST is a key architectural style in enabling interaction with different layers of data. As one of the complex information models, BIM would benefit much from RESTful data interchange, RESTful interactions and RESTful APIs. The following sections will present advanced SOA/RESTful patterns to facilitate interaction with BIMs. The patterns presented in the following sections are RESTful, utilize REST for information interchange, the services presented in the patterns are RESTful APIs and it is assumed that a service consumer API is present and facilitates interaction between the service itself and the service client(s).

6.3   Generalized Design Pattern for BIM-Based SOA

The patterns explained in the following sections define software architectures that consist of several layers. Although the patterns are defined for different purposes, the software layers defined in these patterns have common characteristics. In order to prevent repetition of these characteristics for each pattern, the common character-istics of the architectural layers are illustrated in Fig. 6.1 and are explained below.

1. **Data Layer**: The patterns describe architectures where an extended BIM, a BIM or a BIM view resides in an object or an object relational database (which are shown with database symbols in the pattern illustrations). These databases are the persistence environments where the persistent versions of the BIMs reside. In some patterns the BIM instances are persisted in the form of a BIM file which is encoded as an ISO10303 P-42 ASCII file or an XML file (which are shown as file symbols in pattern illustrations). The databases reside in a database server or in cloud database server hardware. The file resides in a web server or a cloud web server. These data stores together form the first layer of the architecture.
2. **Database/File API Layer**: The database/file API forms the second layer of the architecture. The API acts as a database/file interface that will be used to query and interact with the object/object-relational databases or the file, based on requests coming from the run-time objects (i.e. the upper layer). For IFC BIMs the variations of SDAI-based APIs and XML APIs can be used in this layer. The database/file API can reside within the same hardware which the database

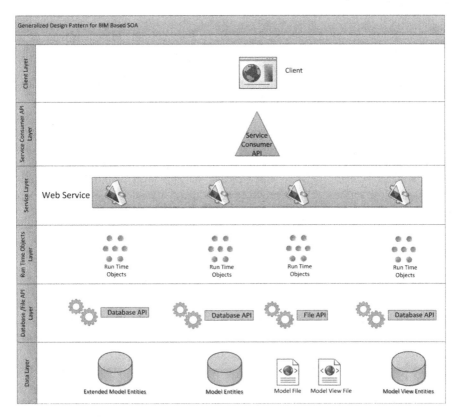

Fig. 6.1 Generalized design pattern diagram

operates (i.e. in the database/cloud database server) or which the file resides in, but it can also reside in a different server.

3. **Run Time Objects Layer**: The run time objects form the third layer of the architecture and maintain the (i) transient data objects that are generated as a result of service requests/interaction and (ii) transient objects of the business (or pattern) logic which the service designer chooses to utilize in this layer. The communication of this layer is bidirectional (i.e. would either be for information provision to the service or for updating the data layer based on information provided by the service) and the direction of communication depends on the type of REQUESTs that are directed to the Web service. The object container can be one of Common Language Infrastructure of the .NET cross-language cross-platform object model, EJB contained within the Java VM. The service designer can choose to distribute the business logic between the service objects and the run time objects. The run time objects layer will be tightly coupled with the database/file APIs and the service. The ratio of objects maintained in the run time objects layer to objects maintained in the service layer would be defined by

the choice of the software architect. Both the run time objects layer (i.e. the object container) and the service layer can be located in the same hardware or in different hardware components/servers, including real and virtual servers. The number of predicated requests/interactions between these layers per time frame (i.e. hour/day/week etc.) and the network bandwidth required for communication between these layers will be the main determinants for this decision.

4. **Service Layer**: This layer consists of service component(s) that provide the RESTful interface. This is the core service layer of the pattern. Each service component presented in this layer will be implemented as a REST API. This layer's main function is to provide an interface to handle HTTP GET/POST/PUT/DELETE REQUESTS, which means that this layer would form the endpoint for the service. All the requests from client side such as

HTTP GET http://www.service.com/model
HTTP POST http://www.service.com/model
HTTP PUT http://www.service.com/model
HTTP DELETE http://www.service.com/model

will be handled by this layer. Based on architectural decisions, this layer can also contain server-side business logic along with service logic to handle HTTP REQUESTs. The service discovery and metadata provision mechanisms are also implemented in this layer. It is advised to maintain a dedicated real or virtual server as the hardware for this layer to operate smoothly.

5. **Service Consumer API**: The service consumer API forms the next layer of the architecture. The API is a general-purpose software component which is loosely coupled with the REST API. The consumer API can reside in an independent hardware, or on the client side. Unlike with the choice in run time objects layer, this layer does not contain the client-side business logic (such as software components for user experience or visualization support). The API here is solely designated to facilitate communication between the client and the REST API. This API acquires information from the REST API and this information is then transferred to the client. The API can also be used to transfer information from the client to the service side. Communication between the service consumer API and the REST API needs to be efficient and most of the bandwidth requirement of the overall architecture would be for this API. If an efficient communication mechanism between these two APIs cannot be established, it can generate a major bottleneck for the overall architecture.

6. **The Client**: The client is the software or a software component that is loosely coupled with the RESTful Web service. In other words, the client would exist and function perfectly without the existence of the Web service; similarly the Web service presented in the patterns would exist and function perfectly without the requirement of the existence of the client. The client can be a simple visualization interface working on a PC, a CAD or analysis software, a Web-based 3D visualization tool, a mobile device/tablet interface, an augmented

or virtual reality device user interface or any other user interface. The client provides the means to the user for interaction with the presented visualization of the model (which can either be in the form of a 3D model or a simple tabular data). The visualization of the information that is acquired from the Web service and the acquisition of the user input are the key functions of the client. Apart from its main functionality and based on user requirements, the client can provide voice or video communication with other clients. Following the generalized design pattern, the specialized patterns developed will be explained in two parts, first a problem definition will be provided, and then the structure of the service pattern and its role in solving the (early defined) problem will be presented.

6.4   REST Query Filter Pattern

The Problem In BIM-based collaborative environments there usually exist multiple stakeholders and multiple synchronous activities. Furthermore, a single user might require access to a set of selected elements either from the BIM (i.e. the model), an extended information model (i.e. extended BIM) or from a BIM view (i.e. model view). Multi-user environments with synchronously working devices can generate a lot of network traffic and require more bandwidth; furthermore, if the models are residing in a cloud database server, transferring the overall model (which might be large to several hundred MBs) for every transaction or user request will also be economically unfeasible.

The Solution As a solution to the above-mentioned problem, the REST query filter pattern describes a software architecture, where information from a set of BIMs, extended BIMs and BIM views can be transferred efficiently to the requesting parties over the Web using discrete and uncoupled query filter services for each data resource. The architecture consists of six layers.

1. **Data Layer**: This layer implements generalized design pattern data layer, which consists of an extended BIM, a BIM or a BIM view.
2. **Database API Layer**: The layer implements generalized design pattern database API layer.
3. **Run Time Objects Layer**: The layer implements generalized design pattern run time objects layer.
4. **Query Service Filter Layer**: This layer consists of discrete service components that provide RESTful interfaces for interacting with different data models of BIMs, extended BIMs and BIM views (Fig. 6.2). This service will focus on providing a filter to present similar building components (i.e. a set of columns, beams, fire alarms) or components of a group of building elements (first-floor elements, façade elements) or elements with time history (i.e. updated after last

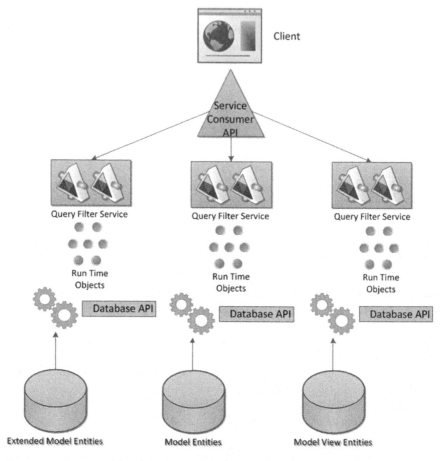

Fig. 6.2 REST query filter

year or updated in the last 100 days). An interaction with these API can include
REQUESTS such as

HTTP GET http://www.service.com/model/filter/beams
HTTP GET http://www.service.com/modelview/filter/firstfloor
HTTP GET http://www.service.com/extendedmodel/filter/façade
HTTP GET http://www.service.com/model/filter/beams/updated/days/100
HTTP GET http://www.service.com/model/filter/walls/updated/years/1

5. **Service Consumer API**: This layer implements generalized design pattern
   service consumer API layer.
6. **The Client**: This layer implements the client description in generalized design
   pattern.

6.5 REST Façade Pattern

The Problem Today, design process in the construction industry requires tools for synchronous collaboration between the stakeholders. For example an architect, an engineer and a customer would have the need to work on the design documents (which are generated from the BIM) synchronously. In fact these parties are mostly located in different places, and it is difficult to make efficient synchronous use of design tools because of their location constraints. Thus there appears a need for collaborative use of information systems over the Web. In fact, as different stakeholders focus on different aspects of the process and need to interact with different parts of the data store, problems occur in reaching information from multiple model (data) sources such as the BIM itself, the extended model or the model view.

The Solution In order to propose a solution to the problem, REST façade pattern provides a single interface to multiple components of the data layer, which will act as a RESTful gateway to reach information in extended BIM, a BIM, a BIM view stored in databases or the model information residing in BIM files. The RESTful architecture provides loose coupling between the client and the provided façade. The architecture consists of six layers.

1. **Data Layer**: This layer implements generalized design pattern data layer, which consists of an extended BIM, a BIM, a BIM view or a BIM file which can be queried through the RESTful façade.
2. **Database/File API Layer**: The layer implements generalized design pattern database/file API layer.
3. **Run Time Objects Layer**: The layer implements generalized design pattern run time objects layer.
4. **REST Façade Layer**: The layer consists of a single service interface. The role of the interface is to provide a service to enable interaction with the multiple data sources through the Web (Fig. 6.3). The service will focus on queries targeted to multiple data sources such as acquiring the garden furniture information from the extended model while acquiring the outer installations of the building from the model (BIM) itself. Another query can be for exploring utilities inside the building together with utility elements in the garden (i.e. which are represented in the extended model). An interaction with the REST API of this service can include REQUESTS such as

   HTTP GET http://www.service.com/façade/outside_elements
   HTTP GET http://www.service.com/façade/all_utilities

5. **Service Consumer API**: This layer implements generalized design pattern service consumer API layer.
6. **The Client**: This layer implements the client description in generalized design pattern.

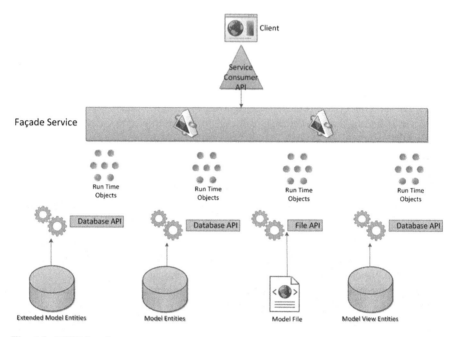

Fig. 6.3 REST façade

6.6   RESTful Real-Time View Generator Pattern

The Problem In multi-stakeholder processes of the building lifecycle, especially during the design and construction process, there is an ongoing need to reach the different elements of the model (related to the role of the stakeholder) on demand, by multiple stakeholders located in different places. The information requests sometimes are not fulfilled by early defined views as they do not cover the aspects for the required information, or transferring an overall model view(s) to the multiple stakeholders would require high bandwidth.

The Solution The RESTful real-time view generator pattern presents a solution to the problem by enabling the generation of on-demand model views that are tailor-made for the user requests. The pattern includes a service that will respond to requests of the users to generate the model views. The RESTful architecture provides loose coupling between the client and the provided service. The architecture consists of six layers (Fig. 6.4).

1. **Data Layer**: This layer implements generalized design pattern data layer, which consists of a BIM or a BIM file which can be queried through the RESTful real-time view generator.
2. **Database/File API Layer**: The layer implements generalized design pattern database/file API layer.

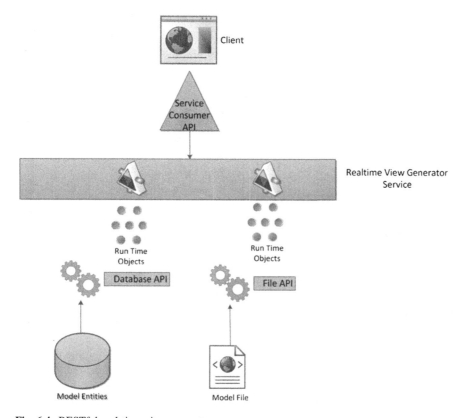

Fig. 6.4 RESTful real-time view generator

3. **Run Time Objects Layer**: The layer implements generalized design pattern run time objects layer.
4. **RESTful Real-Time View Generator Layer**: The layer consists of a single service interface. The role of the interface is to provide a service that will generate transient model views based on the requirements of the users. For example, a mechanical engineer would like to visualize the utility elements or HVAC elements of the second floor, or a civil engineer would like to check the details of columns on the first floor, or the architect would like to check the details of window elements. An interaction with the REST API of this service can include REQUESTS such as

HTTP GET http://www.service.com/runtimeview/utilities/secondfloor
HTTP GET http://www.service.com/runtimeview/columns/firstfloor
HTTP GET http://www.service.com/runtimeview/windows
HTTP PUT http://www.service.com/runtimeview/columns/firstfloor

The service would also be able to handle the HTTP PUT/POST/DELETE REQUESTS; for instance, the last example PUT REQUEST can be used to update the BIM with the as-built information provided from the construction site. In this case, the service (i.e. this layer) would interact with the run time objects layer to update the BIM with the latest changes that occur at the construction site.

5. **Service Consumer API**: This layer implements generalized design pattern service consumer API layer.

6. **The Client**: This layer implements the client description in generalized design pattern.

6.7   RESTful Memento Pattern

The Problem In the design and construction process the BIMs are updated frequently and the stakeholders have access to the latest version of the BIM. As this has been the key user requirement of the BIM-based construction management processes, most efforts focus on providing the most up-to-date version of the model. In fact, in many situations, specifically in the design process and less commonly in the construction process, there appears a need for examining the previous version of the model to compare the changes, and to notify what has changed and so on.

The Solution The RESTful memento pattern is focused on persisting the multiple state(s) of the BIM in the data layer and restoring an old version of the model when required by the user. In other words, the pattern focuses on enabling the backing up of the previous versions of the BIM in a persistence environment and restoring them. The memento service provided in this pattern would (i) generate a back-up copy of the model with the timestamp as a response to a user REQUEST (i.e. on demand) and would store this data in a model server database and (ii) would restore a model from the generated copies based on user REQUEST. The RESTful architecture provides loose coupling between the client and the provided façade. The architecture consists of six layers (Fig. 6.5).

1. **Data Layer**: This layer implements generalized design pattern data layer, which consists of a BIM that resides in a model server database.

2. **Database API Layer**: The layer implements generalized design pattern database API layer.

3. **Run Time Objects Layer**: The layer implements generalized design pattern run time objects layer.

4. **RESTful Memento Layer**: The layer consists of a single service interface. The role of the interface is to provide a service to generate a copy of the overall BIM based on a service call, and store this query in a model server database (i.e. where the current model resides). A BIM data warehouse can be built upon the stored versions of the model in a later stage. An example call to the service would involve a real-time back-up request or a batch request, or a scheduled request which can be accomplished when system resources at the data layer

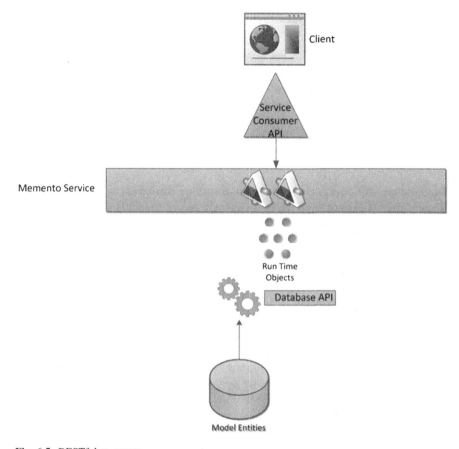

Fig. 6.5 RESTful memento

become free or at the scheduled time intervals. The service would also have the capability to restore the model from one of the previous versions. An interaction with the REST API of this service can include REQUESTS such as

HTTP GET http://www.service.com/memento/backupnow
HTTP GET http://www.service.com/memento/backup/weekly/saturday/22/30
HTTP GET http://www.service.com/memento/backup/monthly
HTTP GET http://www.service.com/memento/restore/version/date/10/20/2015
HTTP GET http://www.service.com/memento/restore/version/last
HTTP GET http://www.service.com/memento/restore/version/lastweek

5. **Service Consumer API**: This layer implements generalized design pattern service consumer API layer.
6. **The Client**: This layer implements the client description in generalized design pattern.

6.8   RESTful Model Multi-view Controller Pattern

The Problem The process in the BIM-based construction management requires environments for supporting synchronous communication and collaboration between the stakeholders. For example, an architect, an engineer and a customer would like to observe updates to the project schedule (on finished tasks which are) maintained by the BIM in real-time (while a site engineer is controlling the processes in a construction site). This process would require real-time updates to all stakeholders' devices/user interfaces regarding the real-time representation of the 4D visualization of the BIM and the project schedule.

The Solution The traditional model-view-controller pattern focuses on decoupling the user interface (view) from the business logic objects (model). A controller layer is located between the user interface and the business logic to update the states of the business logic objects based on the user interaction. In the pattern the views are observers of the model; once the model changes the views get notified. This traditional approach enables automatic update of the views once the model changes. In multi-view architectures the pattern is extremely useful as all changes in the model object's state are automatically reflected in the views. This RESTful model multi-view controller pattern provides an architecture for adopting the MVC approach for facilitating BIM-based collaboration systems, in order to enable real-time representation of 4D information in multiple clients (and user interfaces). Although the pattern has similarities with the traditional MVC there are also some differences. The main difference is at the controller component of this pattern which updates the views, in contrast to the traditional MVC where the views are updated by the model. Thus the business logic to update the views is shared between the model and the controller component. In the presented pattern RESTful MMVC, the number of views is more than one, the controller is implemented as a RESTful service, and the model consists of two layers (one being the run time objects and the other being the persisted information model that resides in the database). The controller role is overloaded in this pattern. The architecture consists of six layers.

1. **Data Layer**: This layer implements generalized design pattern data layer, which consists of a BIM that can be queried through the run time objects.
2. **Database API Layer**: The layer implements generalized design pattern database API layer.
3. **Run Time Objects Layer**: The layer implements generalized design pattern run time objects layer; in addition, components in this layer make calls to the controller service layer to update the view.
4. **Controller Service Layer**: This layer consists of a single service interface. The role of the interface is to provide a controller layer. At the start of the sequence, each client (view) subscribes to the controller service by providing its GUID or IP address in order to get up-to-date information about the model (i.e. both transient objects and persistent model). Once subscribed, each view is provided with the permission to manipulate the model. Upon user interaction with the

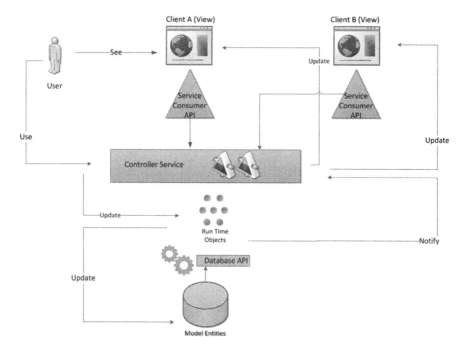

Fig. 6.6 RESTful MMVC

view, the controller service is notified by the service consumer API, which would also provide information regarding what has changed in the view. The controller service once being notified by the call of the service consumer API, notifies the run time objects which then manipulates the transient model objects and the persistent model in the database. Following the persistence of the change (or update) in the data layer, the run time objects interact with the controller service REST API to update the views by sending the changes in the model, for example, as a .json message. The REST API would then interact with the service consumer API to update the views. In this pattern the container of the run time objects would reside in a different platform/or even hardware from the service container (Fig. 6.6). An interaction with the REST API of this service can include REQUESTS such as

Client → Controller HTTP GET http://www.srv.com/controller/subscribe/id/2/ip/22.11.11.22:7777
Service Consumer API → Controller HTTP PUT http://www.srv.com/controller/update/model {data: json message}
Run time Objects → Controller HTTP PUT http://www.srv.com/controller/update/view/2 {data: json message}
Run time Objects → Controller HTTP PUT http://www.srv.com/controller/update/allviews {data: json message}

5. **Service Consumer API**: This layer implements generalized design pattern service consumer API layer.
6. **The Client**: This layer implements the client description in generalized design pattern. In fact there are multiple clients in this layer.

6.9 RESTful Call-Back Responder Pattern

The Problem A standard BIM-based collaboration environment consists of multiple users that are connected concurrently to a model server, either in a tightly coupled manner or in a loosely coupled one (such as over a REST API). In addition to these constraints a REST API in the architecture should need to respond to multiple queries in a sequence without causing latency for the client. In some situations these queries would include time-consuming calculations on the server side (such as structural analysis) which would generate responses that are very late for the client and this would cause inefficiencies in the architecture.

The Solution The term "blocking code" refers to a problem in computer programming where a piece of code would need to wait for a response from an operation in order to move on to the next step. If the operation—that is waited for— takes very long to complete, the overall response time increases. This situation is termed as the blocking code. The blocking code can appear anywhere in a service-oriented architecture (SOA), but if it happens on the server side, all clients will be affected by the poor time performance of the service. The call-back functions are the mechanism designed to tackle the problem of blocking code, and are implemented in some server-side programming frameworks such as JQuery, Node. JS/Express.JS. Operations that take a long time to complete in BIM-based collaborative environments can be facilitated by the use of REST APIs developed to support call-back functions. Frameworks such as Node.Js are ideal for the development of these services. The architecture consists of six layers.

1. **Data Layer**: This layer implements generalized design pattern data layer, which consists of a BIM or a BIM file which can be queried through the RESTful real-time view generator.
2. **Database API Layer**: The layer implements generalized design pattern database API layer.
3. **Run Time Objects Layer**: The layer implements generalized design pattern run time objects layer.
4. **Call-Back Responder Layer**: The layer consists of a single service interface developed with a non-blocking coding framework. The role of the interface is to provide a service that is developed using development environments that support the use of call-back functions. For instance, as illustrated in Fig. 6.7 there can be a REQUEST A to the service which takes 7 s to complete; during this

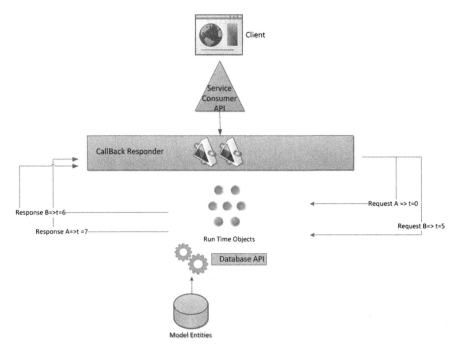

Fig. 6.7   RESTful call-back responder

process there can be another request, i.e. REQUEST B to the service 5 s after the first request. In such a situation as the call-back responder layer does not have to wait for REQUEST A to be completed, it can send REQUEST B to the run time objects layer to be processed. This second process can take 1 s to be completed and response to REQUEST B can be provided in the sixth second, while response to REQUEST A is provided at the seventh second, i.e. after the completion of the second request. REQUEST B in this situation is responded to without any latency. An interaction with the REST API of this service can include REQUESTS such as

HTTP GET http://www.service.com/callbackresponder/makeanalysis
HTTP GET http://www.service.com/callbackresponder/beams/secondfloor

5. **Service Consumer API**: This layer implements generalized design pattern service consumer API layer.
6. **The Client**: This layer implements the client description in generalized design pattern.

6.10   RESTful Authenticator Pattern

The Problem In multi-user environments for collaboration, such as systems used for BIM-based management of the building design and construction, there will be many stakeholders that require access to the BIM Web service. In fact, access by unauthorized parties would create chaotic situations as the models can be updated with unwanted information, or key information regarding construction process can be stolen or deleted.

The Solution This pattern presents an authentication architecture that is loosely coupled with the service. The architecture for the RESTful service that would interact with the model (i.e. BIM) is the architecture presented in the generalized design pattern, in fact this pattern has additional layers. An additional data layer and an additional service layer are present to serve for authentication purposes. The architecture in principle focuses on disabling the service discovery of the RESTful service that would interact with the model. In fact, the URI of the RESTful service will be provided at a later stage by the authenticator service as a result of successful authentication of the client side (Fig. 6.8). Further information on SOA authentication patterns can be found in Erl (2009). The architecture consists of six layers.

1. **Data Layer**: This layer implements generalized design pattern data layer that consists of a BIM or a BIM file which can be queried through the RESTful

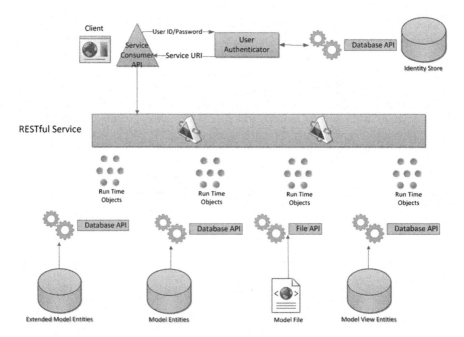

Fig. 6.8 RESTful authenticator

real-time view generator. This layer also implements another data layer where a relational or object relational database/cloud database (i.e. the identity store) holds the user information which will serve for client authentication.

2. **Database/File API Layer**: The layer implements generalized design pattern database API layer. The layer consists of another database API to enable interaction between user and authenticator service.

3. **Run Time Objects Layer**: The layer implements generalized design pattern run time objects layer. The layer does not exist on the authenticator side.

4. **User Authenticator Layer**: The layer consists of two service interfaces (i.e. user authenticator, and RESTful service). The user authenticator service will operate as an authentication gateway. A client will call the service with an authentication request by providing a username or a password. The service will then query the identity store and if the information matches the records in the identity store, it will respond to the URL of the RESTful service that the client requested. The user authenticator service can be developed in such a way that it will provide a set of URIs as a result of successful authentication. An interaction with the REST API of this service will include a REQUEST such as

   HTTP POST http://www.service.com/authenticator {data: json message}

5. **Service Consumer API**: This is a generalized API but also implements a function for passing the URI of the RESTful service (sent by the user authenticator) to the client.

6. **The Client**: Once the client finalizes the authentication and acquires the URI of the REST API of the target service it implements the client description in generalized design pattern.

6.11   RESTful Data Management Pattern

The Problem Interactions with building information models do not only involve data acquisition or data update functions at the fine granular object level. Sometimes the data layer might be complex and fuzzy in distributed systems. For instance, BIMs in use might not reside in a single database server, and the project may not contain a single standard BIM such as an IFC, but multiple BIMs. There might be different models defined with different schemas (i.e. Green Building XML, CIS2, IFC ...) which reside in different platforms. In these situations, management and housekeeping at the back-end (i.e. the data layer) become vitally important. If the back-end is not managed successfully the problems will lead to chaos where information exchange and sharing would become a nightmare with so many models defined with different schemas, conflicting views, files that are not persisted in model servers and different versions of different models independently floating around in the data layer.

The Solution The RESTful data management pattern introduces a set of Web services to facilitate data management tasks in a distributed system. The Web services introduced in the pattern can be thought of as a Swiss Army knife for managing the information transformation tasks and facilitating data persistence in model servers. Two of the Web services concentrate on transforming information between two information models, while the other concentrates on persisting the BIM file contents (which has arrived into the data layer as a result of data exchange) in model server databases. The main difference in the architecture presented here from other patterns presented in this section is that the client layer includes data components (i.e. which can be regarded as the target data layer). The architecture consists of eight layers.

1. **(Source) Data Layer**: This layer is shown at the bottom of the diagram and implements generalized design pattern data layer, which consists of a BIM or a BIM file which can be queried through the Web services defined in this pattern.
2. **(Source) Database/File API Layer**: The layer is shown at the bottom of the diagram and implements generalized design pattern database/file API layer
3. **(Source) Run Time Objects Layer**: The layer is shown at the bottom of the diagram and implements generalized design pattern run time objects layer (Fig. 6.9)
4. **Model Management Service Layer**: The layer consists of three different service interfaces. The first one is the model transformer service which can be used

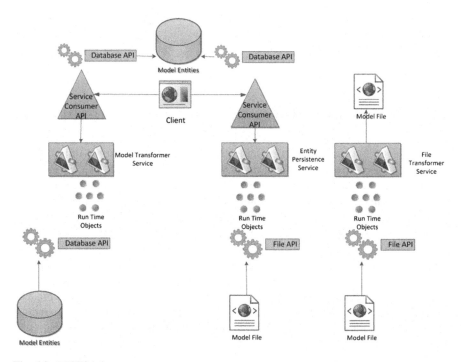

Fig. 6.9 RESTful data management

to transform a source information model with a different schema to the target information model (which is the main data source of another business process). The second service is the entity persistence service which can be used to transfer model entities residing in a file to the model server database in the target data layer. The third service is the file transformer service, which can be used to transform an exchange file of a BIM formatted with source schema to exchange file of a BIM formatted according to the rules of a target schema. For example, a transformation from IFC BIM to CIS/2 BIM can be accomplished using the service layer defined in this pattern. A request from the client to the service starts the transformation process, and the transformed model entities would then be persisted either (i) in target data layer databases using the service consumer API and the database API of the target data layer or (ii) in a target BIM exchange file. An interaction with the REST API of this service can include REQUESTS such as

HTTP GET http://www.service.com/modeltransformer/sourcemodel_id/001/targetschema/ifc2x4
HTTP GET http://www.service.com/entitiypersistence/sourcemodel_id/002
HTTP GET http://www.service.com/filetransformer/sourcemodel_id/001/target schema/ifc2x4

5. **Service Consumer API**: This layer implements generalized design pattern service consumer API layer.
6. **The Client**: Client is responsible for providing the user interface to manage source/target files and databases, and managing the transformation operations by issuing call to the REST API of the model management service layer.
7. **(Target) Database API**: The layer implements generalized design pattern database API layer for the target database.
8. **(Target) Data Layer**: This layer implements generalized design pattern data layer where a BIM is persisted in an object database and in an exchange file.

6.12   RESTful View Synchronizer Pattern

The Problem In multi-user multi model distributed system architectures, which can be beneficial for facilitating the communication, interoperability and information, the clients (i.e. the user interfaces) would have a need to combine information acquired from multiple Web services. In fact as these services are also endpoints which are loosely coupled with the client, the data transfer speed from each different service can vary depending on the location of the servers' bandwidth in the communication process. This variance will result in latency for visualization of information when some services wait for other services to respond and send information. This would create a blocking situation for the visualization environment, which prevent timely visualization of information sent by the Web services.

The Solution The pattern view synchronizer explained here is adopted from the UI mediator pattern (Erl 2009). The architecture of this pattern utilizes a view synchronizer service which will act as a component that will synchronize information coming from multiple sources. The synchronizer service presented in this pattern has two functions. It synchronizes the information sent from multiple endpoints (REST APIs) and also acts as a façade layer and provides a single gateway to multiple REST endpoints. The architecture consists of seven layers (Fig. 6.10).

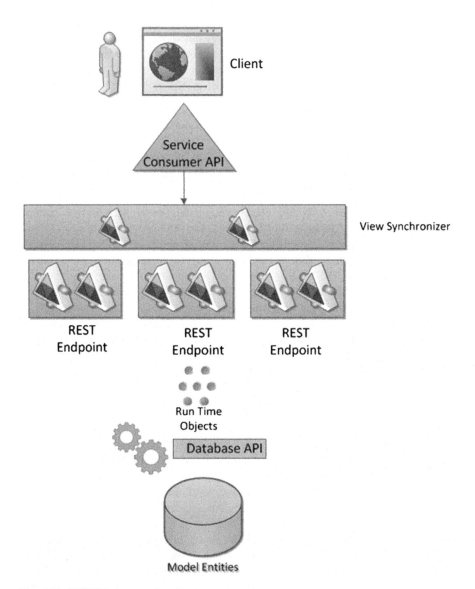

Fig. 6.10 RESTful view synchronizer

1. **Data Layer**: This layer implements generalized design pattern data layer, which consists of a BIM residing in a model server database.
2. **Database API Layer**: The layer implements generalized design pattern database API layer.
3. **Run Time Objects Layer**: The layer implements generalized design pattern run time objects layer.
4. **REST Endpoint Layer**: The layer consists of a multiple service interfaces which comply with the definitions in generalized design pattern.
5. **View Synchronizer Service Layer**: The layer consists of a single service interface. The role of the interface is to provide a service that provides the synchronized information derived from multiple REST Endpoints (i.e. REST APIs). As the business logic for the synchronization would be towards enabling automatic synchronization, the interaction with this layer would be similar to the REST façade pattern. An interaction with the REST API of this service can include REQUESTS such as

   HTTP GET http://www.service.com/synchronized/outside_elements
   HTTP GET http://www.service.com/synchronized/all_utilities

6. **Service Consumer API**: This layer implements generalized design pattern service consumer API layer.
7. **The Client**: This layer implements the client description in generalized design pattern.

6.13   RESTful Event Manager Pattern

The Problem As explained in Chap. 2, building information modelling in the near future would require real-time information provision regarding the states of the elements. This information will be represented in the BIM in real-time. For instance, a state of the door (such as being open or shut) or a state of the elevator (being busy or idle) would reside in BIM databases to support smart city and smart building applications. The next chapter will provide patterns for information provision from sensors along with BIM for supporting these applications. The information consumers in this case are not interested in the information unless a state change occurs. In other words, if the client is notified when a door's state has changed from open to closed, the client is not interested in updates on the state of the door every minute, but would like to get notified when the door's stage changes back to open again. In fact, the Web services are mainly capable of providing the information as a response to user request but not when a change occurs on the server-side model.

The Solution In a situation where a view (client) interacts with the model to change its state it would be feasible to implement the RESTful MMVC pattern; in fact, in this situation the role of the client is only visualization, not interaction with the service. Erl (2009) proposes the event-driven messaging pattern for messaging services for a similar situation. The idea of the event manager proposed by Erl (2009) can be extended to define an event manager service, which is working with publish/subscribe approach. Once the client subscribes for specific events, the event manager can send notifications to the client once the event occurs (such as the state change of door from open to close).

1. **Data Layer**: This layer implements generalized design pattern data layer, which consists of a BIM residing in a model server database
2. **Database API Layer**: The layer implements generalized design pattern database API layer.
3. **Run Time Objects Layer**: The layer implements generalized design pattern run time objects layer; in addition, components in this layer make calls to the controller service layer to update the client.
4. **Event Manager Layer**: The layer consists of a single service interface. At the start of the sequence each client (view) subscribes to the event manager service by providing its GUID or IP address in order to get state changes of the model. Once a state change is represented in the model (such as the start of air conditioning units in the building), the run time objects interact with the event manager to update the views by sending the changes in the model, for example, as a .json message. The REST API would then interact with the service consumer API to update the client. In this pattern the container of the run time objects would reside in a different platform/or even hardware than the service container (Fig. 6.11). An interaction with the REST API of this service can include REQUESTS such as

   Client → Controller HTTP GET http://www.srv.com/emanager/subscribe/id/2/ip/22.11.11.22:7777
   Run time Objects → Event manager → Service Consumer API
   HTTP PUT http://www.srv.com/emanager/update/client {data: json message}

5. **Service Consumer API**: This layer implements generalized design pattern service consumer API layer.
6. **The Client**: This layer implements the client description in generalized design pattern.

This chapter provided software architectures for provision of BIM information through Web service. The following chapter will elaborate on software architectures for provision of sensor information that is acquired from indoor sensors.

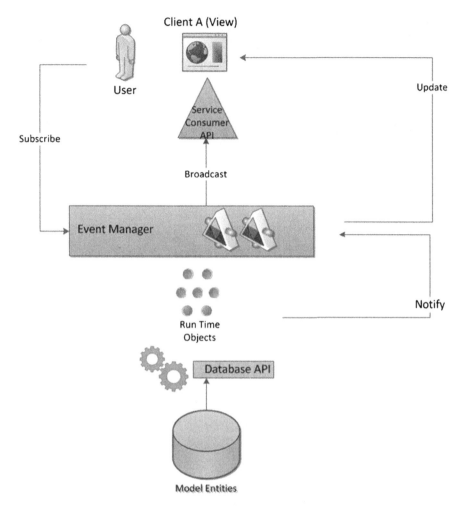

Fig. 6.11 RESTful event manager

References

Erl, T.: SOA Design Patterns. Prentice Hall, New Jersey (2009)

Fielding, R.T.: Architectural styles and the design of network-based software architectures. Ph.D. thesis, Department of information and computer science, University of California, Irvine (2000)

He, H.: What is service-oriented architecture. Online at http://webservices.xml.com/pub/a/ws/2003/09/30/soa.html. Accessed 21 July 2004 (2003)

Isikdag, U., Underwood, J.: Two BIM based web-service patterns: BIM SOAP façade and RESTful BIM, construction in the 21st century conference, Istanbul, May 2009 (2009)

Pautasso, C., Zimmermann, O., Leymann, F.: Restful web services vs. "big" "web services: making the right architectural decision" WWW '08: proceeding of the 17th international conference on World Wide Web, pp. 805–814 (2008)
Pulier, E., Taylor, H.: Understanding Enterprise SOA. Manning Publications, Greenwich (2006)
RESTful API Tutorial: http://www.restapitutorial.com/lessons /restfulresourcenaming.html (2015)
Techtarget: Definition of REST available at : http://searchsoa.techtarget.com/definition/REST (2015)

Chapter 7
Sensor Service Architectures for BIM Environments

Abstract The Internet of Things (IoT) approach proposes a global wireless sensor/actuator network composed of everyday devices such as home appliances, city furniture, mobile phones or vehicles. Everyday devices would be either publishers of information and subscribers of information coming from other people and devices. The information provided by each device would only be limited by the device's sensing capabilities. In order to represent real-time information about indoors, the integration between information acquired from IoT devices and BIM elements (including door, window, wall, slab, roof, staircase and space) is an obvious necessity. This chapter first presents patterns related to integration of information provided by IoT elements, and then focuses on patterns that are formulated for integration of BIMs with information provided by the IoT elements.

7.1 Introduction

IBM's Smarter Planet Video (ASmarterPlanet 2010) starts with the following statements:

> Over the past century, but accelerating over the past couple of decades, we have seen the emergence of a kind of global data field, the planet itself. Natural systems, human systems and physical objects have always generated an enormous amount of data, but we did not use to be able to hear it, to see it, to capture it. Now we can, because all of the stuff is instrumented, and it is all interconnected, so we can actually have an access to it, so in effect, the planet has grown a central nervous system....Over the last 10 years devices are being linked up together using networks, such as temperature sensors, flow rate sensors, electricity measuring devices, and it will not be long or it may even have happened already that, there are more things on the Internet than there are people on the Internet. That is really what we mean by the Internet-of-Things

Ubiquitous computing and the Internet of Things (IoT) concepts are gathering more attention day-by-day. From the perspective of information management, realization of ubiquitous computing would lead to a focus shift (in data management) from data acquisition to abstraction of acquired data, as in the future data will

U. Isikdag, *Enhanced Building Information Models*,
SpringerBriefs in Computer Science, DOI 10.1007/978-3-319-21825-0_7

be constantly provided by networks of interconnected sensors monitoring the physical environment around us. This newly emerging paradigm can be referred to as "Sensors Everywhere". This new paradigm will have implications on the management of urban information, where the focus will shift towards the provision and management of real-time city level information.

Sensors can be defined as hardware components that are used to measure various physical quantities such as temperature, humidity, flow rate, force, pressure, gas ratios and so on. The sensors can either output a voltage value (i.e. analog sensors) or a Boolean value (i.e. digital sensors). A sensor network is a collection of sensors that can communicate autonomously with each other or with a central computer in the network. In a sensor network, the nodes can be connected to each other by wires or wirelessly. In the near future, increasing number of connected devices will make use of sensors to acquire information from their surrounding environment. The connected devices (which are also referred as things) will have an ability to interact with other devices over the Internet, provide information in interoperable form and consume/utilize such information. This overall concept is known as IoT. The IoT approach proposes a global wireless sensor/actuator network composed of everyday devices such as home appliances, city furniture, mobile phones or vehicles. These devices would be either publishers of information or subscribers of information coming from other people and devices. The information provided by each device would only be limited by the device's sensing capabilities. Recent research has shown that there is a growing interest in establishing wire and wireless sensor networks indoors to monitor the different characteristics of a building such as energy consumption, indoor air quality, temperature and so on (Kwon et al. 2007; Tse and Chan 2008; Campa et al. 2011).

Recent research has also demonstrated that semantically rich digital models of buildings (which also provide advanced 3D representations of building elements) form the key components of the digital city. Unlike early CAD representations of buildings, in recent semantically rich representations of buildings, the information regarding the building elements is meaningful (i.e. computers can understand a rectangular prism visualized in a model which is not only a simple rectangular box, but represents a room that is in a building used for residential purposes). Recent research and developments have shown that sensors, sensor networks and IoT nodes can be used to collect and present information regarding the state of building elements (such as being open for a window) and indoor environmental conditions. Once provided by the IoT nodes, this information can be used together with the geometric/semantic building information for facilitating many tasks ranging from emergency response to energy management.

Web services was the popular paradigm of the past 10 years. First, Simple Object Access Protocol (SOAP) Web services provided message-based method invocation over the Internet. As most research/studies were focused on user–web or user–software interaction in the early days, heavyweight WS-* technologies (such as SOAP, WSDL, UDDI) worked smoothly. In fact, as mentioned by Guinard et al. (2011) these technologies were too heavy and complex for interacting with the connected devices (i.e. Things). In fact, RESTful architectural style provided significant advantages in (i) facilitating the interaction with IoT nodes and (ii) consuming the information provided by IoT nodes. RESTful architectures were elaborated in Chap. 6.

IoT nodes today are capable of providing real-time information about them on the Web. The information can be acquired from various sensors that are connected to the IoT device and broadcasted through single-board computers (SBCs). Information broadcasted by SBCs is usually in the form of XML or JSON documents. Several RESTful loosely coupled web architectures can be designed to reach and utilize this information. In addition, Usländer et al. (2011) indicated that virtual sensors (i.e. soft-sensors that are used to gather and abstract data from diverse sets of sensor network nodes) can act as a middleware between the sensors and services in these architectures.

The client side of the architecture as indicated in Usländer et al. (2010) can be composed of visualization, reporting and other sensor applications. Since SBCs in an IoT architecture generally have fewer resources than clients, browsers or mobile phones have proven to be a good way of transferring some of the server workload to the client (Guinard et al. 2011), and thus development of rich client applications is also a possibility.

7.2   Sensor and BIM Integration Patterns

In order to represent real-time information about indoors, integration between information acquired from IoT nodes and BIM elements (including door, window, wall, slab, roof, staircase and space) is an obvious necessity. In the first stage, information acquired from the IoT nodes needs to be managed efficiently. This information would need to be integrated with information residing in the BIM. This section will start with providing patterns on exchange of information between the IoT nodes, and storage and presentation of information provided by the nodes, and then focus on patterns that are formulated for integration of BIMs with information provided by the IoT nodes. It should be noted by the reader that the patterns presented in this chapter are mainly focused towards information acquisition and presentation rather than two-way interaction.

7.3   Foundational Publish-Subscribe

The Problem: The IoT approach mainly depends on machine to machine (M2M) communication. In fact, Web service technologies and architectures such as REST generally provide mechanisms for enabling communication based on client actions (and method invocation requests). Thus, a pull approach is common for information acquisition using Web services. In fact, in an IoT environment where thousands of

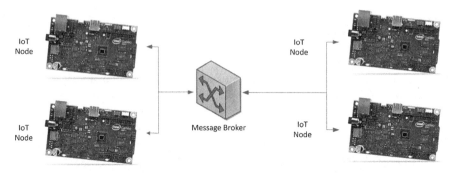

Fig. 7.1 Foundational publish-subscribe. Grant of permission by Intel Corporation

devices connect to each other and exchange information sometimes with very short intervals, a pull-based mechanism is not efficient (Fig. 7.1).

The Solution: The publish-subscribe approach as explained in Chap. 5 forms the backbone of the IoT middleware as it provides an efficient mechanism for devices to share information between each other. Messaging protocols such as CoAP, XMPP and MQTT form the main elements of this approach. The foundational publish-subscribe pattern illustrates a typical architecture to enable and facilitate M2M communication in an IoT environment. In this pattern a message broker acts as a mediator between the IoT nodes. The IoT nodes (composed of single-board computers—SBCs) publish messages to a message broker (such as an MQTT server/broker), which will then distribute this messages to the other IoT nodes which subscribe to the message broker. Protocols such as MQTT and MQTT message brokers are today in use with Arduino devices to facilitate M2M communication in home automation. For instance, a luminosity sensor attached to an Arduino SBC publishes a message to an MQTT broker when the light level in a room decreases below a certain threshold value, then this message is distributed to another SBC which has actuators to turn on the home lighting. Once this message is received by the SBC it activates the lights in the room. This is a basic example of how traditional publish-subscribe works for IoT nodes, but the approach is simple, the protocols in the approach are lightweight and very efficient in enabling device to device (M2M) communication.

7.4   Feed Encoder

The Problem As explained in the previous pattern, message brokers are known as efficient middleware components for distributing messages to other devices; in fact, messages distributed by the brokers can also be consumed by everyday applications, web portals and mobile devices. Although the information acquired from most of the home automation sensors would not be confidential in general, there are

Fig. 7.2 Feed encoder. Grant
of permission by Intel
Corporation

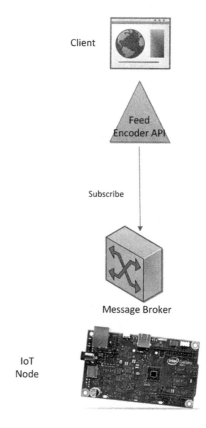

applications concerned with critical sensors where secure transfer of information is
a necessity (Fig. 7.2).

The Solution   The feed encoder pattern provides a mechanism for securely
transferring messages acquired from sensors and IoT nodes to the requesting parties
and information system components. The patterns describe an architecture composed of an IoT node, message broker middleware, a feed encoder API and a client.
The IoT node is formed by an SBC. The feed encoder API is a specific middleware
component that aims to encrypt messages coming from the message broker using
the public key encryption. The information flow sequence is as follows. Once the
information is acquired from the sensors by the IoT node (SBC), it is first encrypted
using a symmetric key cryptography technique.The encoded message is published
to the message broker and received by the feed encoder API, which is a subscriber
of the message broker. If the message coming from the message broker is
encrypted, the feed encoder API will decrypt the message using the shared symmetric key and re-encrypt it using a public key cryptography technique. The
(second) encrypted message would then be transferred to the client, which will then
decrypt it using the public key (asymmetric) cryptography. In addition to this

technique, the use of public key authentication between the IoT node, message broker and feed encoder API and sending the message to the feed encoder API without encryption is the other alternative.

7.5 Message-Based Cloud Update

The Problem   The IoT nodes publish a large amount of data in a short time frame. The use of real-time data coming from sensors is important for other devices in M2M communication. In fact the novel paradigm of big data advocates that information acquired regarding the states of everyday objects including city objects, buildings and indoor objects is of key importance and needs to be analyzed to find patterns of behaviours and patterns of occurrences in the cities. Thus, in order to enable this analysis, information acquired needs to be persisted (stored) in databases (Fig. 7.3).

The Solution   As the information stored needs to be accessed from multiple resources for big data analysis, online storage of the data is necessary. In order to store information acquired from the IoT nodes, cloud storage is an economically feasible alternative. The message-based cloud update pattern proposes a six-layer architecture. The first two layers of the architecture are formed by the IoT node and a message broker. The third layer is composed of a message consumer API, which can be defined as a general-purpose API which is the subscriber of the message broker middleware. Once a message is received from the IoT node, the message broker distributes it to the message consumer API. The second role of the message consumer API in this pattern is automatically updating the cloud database/files through the REST endpoint once it receives a message from the message broker middleware. As the service consumer API can be the subscriber of different brokers and brokers might be the observers of many IoT nodes, the bandwidth between the IoT nodes/message broker and message broker/cloud DB REST endpoint should be high. The next layer in the stack is the REST endpoint. The REST endpoint is the general-purpose endpoint to interact with the database API or the file in the cloud layer. The REST endpoint will contain business logic for the I/O operations regarding the file and will also have the ability to communicate with the database API. Although a database API is mentioned as a component in this architecture, it is optional and the REST endpoint can contain business logic to interact directly with the cloud DB. The database API layer contains the database API, an optional component which aims to facilitate the CRUD operations related with the cloud DB. As some cloud DBs in use today provide RESTful endpoints, the use of a specific database API is not always compulsory, but this layer can be required when a special-purpose database (with no RESTful interface) is planned to be used in the

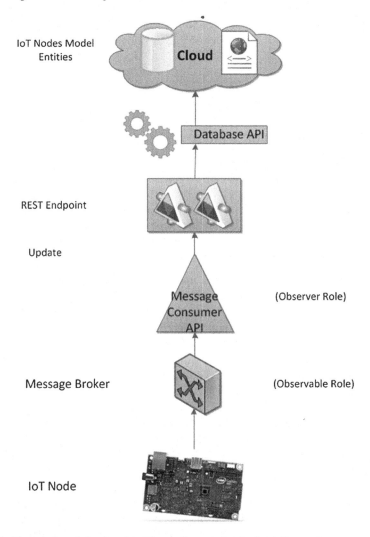

Fig. 7.3 Message-based cloud update. Grant of permission by Intel Corporation

data layer. Finally, the data layer consists of a cloud DB and/or a file. The cloud DB preference in this layer would be a spatial database, graph database or an XML database as spatial, graph and XML databases are capable of storing and presenting information coming from IoT nodes with greater efficiency and with geo-coordinates. The second reason behind this choice is spatial and graph databases would provide a data structure that is more efficient to respond to spatial queries, while XML databases provide structured information that would facilitate big data analysis. The file storage would either contain an XML file or a plain-text CSV (comma separated values) file. As mentioned, XML files provide structured information for big data analysis and plain text files are efficient forms of storage in terms of file size.

7.6   On-Demand Cloud Update

The Problem   The previous pattern message-based cloud update provides an architecture where information acquired from IoT nodes is periodically updated to the data storage components. Depending on the density of the data and frequency of updates, huge amounts of data can be persisted in the cloud DBs in a very short time. This might generate a bottleneck related to the storage space or if not, high data storage cost can appear as an issue.

The Solution   The on-demand cloud update pattern proposes a software architecture for storing the information acquired from the IoT nodes to cloud DBs, only when required by the user. In contrast to the previous pattern where there is a continuous update of data layer components, in this architecture the update of data layer is done only when requested. There are seven layers in the proposed architecture. The first layer is the IoT node which is capable of publishing information acquired from the sensors to an HTTP endpoint. SBCs such as Arduino and Raspberry Pi provide Web servers where a web page can be generated dynamically based on sensor data and stored/served from these hardware components. The second layer is composed of an HTTP endpoint which contains an XML, or HTML file residing in the IoT node (SBC). The XML or HTML file in this layer would be defined as plain/simple as possible. Text files can also be the components of this layer and would contain information as plain text or in CSV form. The role of the REST endpoint is to acquire information from the HTTP endpoint and provide an interface to serve this information to the consumers of this information. The service consumer API is of key importance in this pattern as applications or GUIs will interact with the service consumer API to send the information update requests. Once such a request is received by the service consumer API, the component issues a call to the REST endpoint and acquires information from the IoT node regarding the states of objects that are observed. Later in the next stage, the service consumer API interacts with the REST endpoint of the cloud DB/cloud data layer to update the data layer with the information acquired from the sensors. Similar to the message-based cloud update pattern, the REST endpoint of the cloud DB/cloud data layer is a general-purpose endpoint that provides CRUD functionality for the cloud data layer; the database API is an optional component for bridging the REST endpoint and cloud data layer and the data layer is composed of a cloud DB (which can be a spatial/graph or XML database) and/or an XML/plain text file (Fig. 7.4).

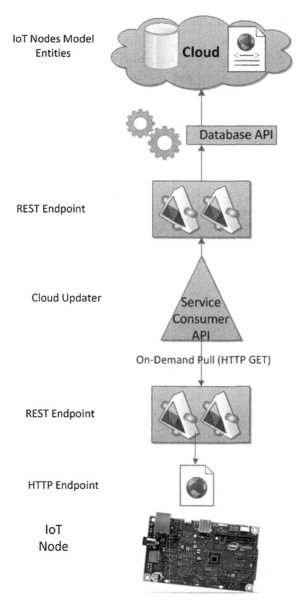

Fig. 7.4 On-demand cloud update. Grant of permission by Intel Corporation

7.7   RESTful Node Façade

The Problem   In the case of direct acquisition of information from multiple IoT nodes, communication of clients with the IoT nodes can be enabled by use of message brokers. A subscription to a message broker can be enabled over an API, but if there are many IoT nodes in the architecture or the IoT nodes are publishing data too frequently, the API will be flooded by messages coming from multiple sensors. In this situation a periodic pull approach needs to be implemented in order to limit the data transfer from the IoT nodes (Fig. 7.5).

The Solution   In order to provide a mechanism to regulate the information transfer from the IoT nodes to the client side, a periodic pull approach is implemented in the RESTful node façade pattern. The pattern presents an architecture composed of five layers. The first layer is composed of IoT nodes. Similar to the previously presented pattern, in this pattern the IoT nodes publish information acquired from sensors to HTTP endpoints. The HTTP endpoints contain simple XML, HTML or text files. The IoT node façade is a RESTful service layer, the service in this layer needs to contain programming logic to acquire information from multiple documents in the HTTP endpoint layer (i.e. fulfilling the façade role); in addition, the IoT node façade service needs to be able to handle periodic pull (HTTP GET) requests from the upper layer. The service consumer API issues periodic pull (HTTP GET)

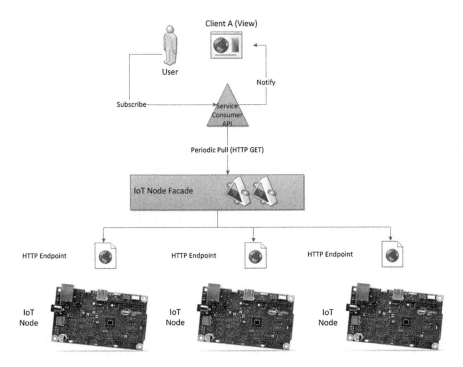

Fig. 7.5 RESTful node façade. Grant of permission by Intel Corporation

requests based on user demand and user–client interaction. The client (which is the subscriber of the service consumer API) provides a GUI to the user, which can also contain business logic to mash-up information coming from multiple IoT nodes.

7.8 BIM and IoT Service Façade

The Problem Novel developments in management of urban and building information indicate the importance of provision of real-time information regarding the states of the objects. Information infrastructure of urban and building information systems mainly depend on models defined compliant with schema standards such as BIMs and CityGML. These models are mainly successful in provision of semantic information regarding the building (elements, interiors and objects indoors) and city objects; in fact, real-time information about the states of these objects is acquired from sensors and SBCs but is not presented in an integrated form with the BIMs and city models.

The Solution The BIM and IoT service façade pattern propose a 10-layer architecture to provide a generalized loosely coupled approach (i) to reach information residing in the BIMs and (ii) to utilize information stored as a result of information acquisition from IoT nodes. The pattern provides a generalized architecture of multi-source information fusion using a RESTful service façade. The IoT node layer is the first layer of the architecture and contains multiple IoT nodes (SBCs) which publish information to a message broker in the second layer. The message broker is in the second layer of the architecture and is responsible for provision of messages to its subscribers. A general-purpose message consumer API forms the next layer, is a subscriber of the message broker and is responsible for updating the upper layer components based on information provided by the message broker. The following API layer consists of database API and file API for IoT model storage components; once updated by the general-purpose message consumer APIs, they interact with the related data layer components in order to update the IoT nodes model and IoT nodes file storage. The data layer consists of the BIM, extended BIM and model view entities stored in model server databases and in file storage. IoT entities stored in NOSQL databases and IoT device files are the other components of this data layer. Different proprietary and shared APIs (some of which are compulsory) interact with the data storage components to perform CRUD operations that are requested by the run time objects. The run time objects layer is formed with a set of software components designed to bridge the façade service and the database layer. The service layer is composed of a general RESTful façade service which aims to enable and facilitate Web interactions with the IoT node models and BIM simultaneously, and carries the key responsibility of enabling information fusion between the BIM and IoT node models. The service consumer API is a general-purpose RESTful API that facilitates the interactions with the RESTful façade service. The client component forms the top layer of the architecture and a

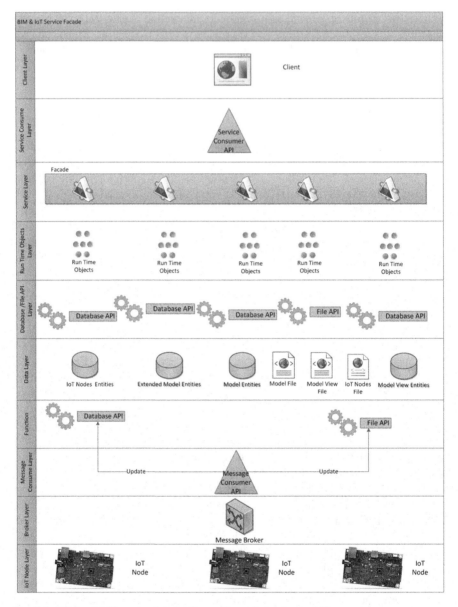

Fig. 7.6 BIM and IoT service façade. Grant of permission by Intel Corporation

client can be a visualization or analysis software that consumes information from
both BIM and IoT nodes. Software such as virtual globes or web portals are good
candidates for the client role of this pattern (Fig. 7.6).

7.9 BIM Updater Nodes

The Problem BIMs of today provide semantic information regarding building elements and indoor spaces, but in fact they are not able to model and reflect the indoor object states (i.e. what is happening indoors?) and indoor conditions. In fact, the BIMs as explained previously can be extended to represent the state information related to indoors. In this case the BIM becomes a real-time information model. Thus, in this situation real-time information acquired from the sensors and BIM entities need to be updated with this information, in order for the BIM to represent real-time information along with the building semantics (Fig. 7.7).

The Solution The BIM updater nodes pattern presents an architecture that would enable the update of BIM entities based on information acquired from the multiple distributed IoT nodes. The pattern has six layers. The first layer is composed of IoT nodes which publish their states as messages to the message broker. The message broker is the subscriber of the IoT nodes and publishes messages to the compound API, which consumes messages from the message broker and based on interaction with the IoT nodes communicates with the RESTful endpoint (i.e. Web service) of the BIM data layer to update the BIM with information coming from IoT nodes. The RESTful endpoint is the general-purpose RESTful Web service designed to facilitate the CRUD operations in the BIM data layer. The RESTful endpoint interacts with the database API of the model server to update the model server

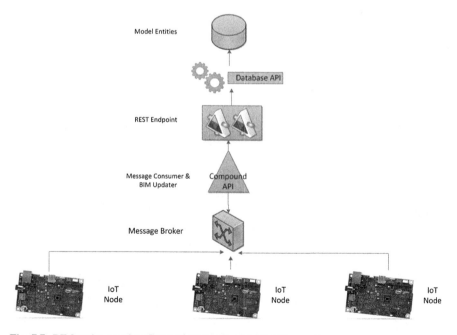

Fig. 7.7 BIM updater nodes. Grant of permission by Intel Corporation

database. Model server database forms the data layer of the architecture. The real-time BIM resides in the model server database which forms the data layer of the architecture.

7.10   Rich Client for BIM and IoT Nodes

The Problem   The softwares that consume information from BIM today are either CAD packages for design or software for analysis of structures. As these tools work with models of non-existent buildings, the need for real-time information from the BIMs is not very apparent currently during design. In fact, city modelling and management applications including city portals in the near future will utilize information acquired from BIMs. These applications and portals will also have access to mash-ups of information published by IoT nodes. In this situation, building information and information regarding the current state of objects indoors and outdoors need to be integrated in these applications and portals (Fig. 7.8).

The Solution   The pattern elaborated in this section proposes the use of rich clients which have the capability of interacting with multiple APIs of BIM and IoT nodes. The architecture described consists of six layers. The IoT nodes publish information to a message broker, and a general-purpose message consumer API serves this information on demand to the rich client. On the other dimension, BIMs (model

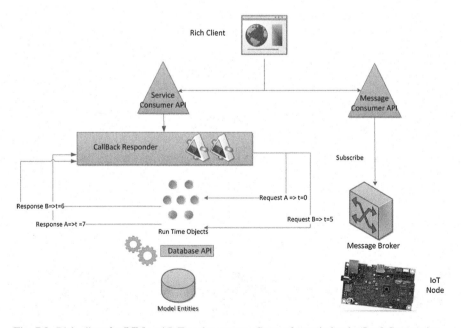

Fig. 7.8 Rich client for BIM and IoT nodes pattern. Grant of permission by Intel Corporation

entities) reside in model server databases; the database API is used to bridge the gap between the run time objects and the data layer. A callback responder service is a RESTful service that supports asynchronous communication with the BIM through run time objects and database API. As there can be many BIMs in the architecture, the run-time efficiency (time performance) is supported by use of callback functions in the service layer. On the other hand, the use of message broker and lightweight messaging protocols such as MQTT will also contribute to the time performance of the system. The service consumer API is a general-purpose API that the rich client will utilize to interact with the callback responder service.

7.11   Real-Time BIM Callback

The Problem   The previous pattern provides a rich client-style solution to service-based integration of BIMs and IoT nodes. In fact, when the client has to discover and connect many BIMs and IoT nodes for acquiring indoor information, it would be problematic when there is no directory service and discovery mechanism (Fig. 7.9).

The Solution   In the situation explained here, a solution to this problem of discoverability would be reducing the information endpoints, updating BIMs with

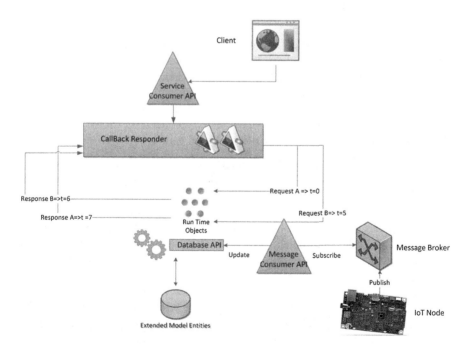

Fig. 7.9   Real-time BIM callback pattern. Grant of permission by Intel Corporation

information coming from the IoT nodes and thus integrating BIMs and IoT nodes at a lower layer and providing a BIM-based single entry point to present indoor information. The real-time BIM callback pattern presented in this section provides a mechanism for such an integration and consists of six layers. The IoT nodes publish information as a result of state changes to a message broker. In this case, a message consumer API is the subscriber of the message broker and is also responsible for interacting with the database API for updating the BIM with information coming from IoT nodes. The database API is also responsible for publishing real-time information from BIM (on-demand) when requested by the client application. The run time objects interact with the BIM to acquire information and present it through the callback service. The callback responder is a RESTful service designed to respond queries in an asynchronous manner. The asynchronous nature of the RESTful service will contribute to the time performance of the architecture. The service consumer API interacts with the callback responder service as result of the request that is coming from the client. The client can be a web portal component or an application that presents information about building interiors or an application focused on real-time city level analysis (such as an application for calculating real-time energy consumption of a city).

7.12 BIM Virtual Sensors

The Problem The communication protocols that enable device to device (M2M) communication form the backbone of IoT, and these protocols mainly utilize message-oriented middleware, such as message brokers, to facilitate exchange of small amounts of data in a very timely manner. In fact, BIM applications utilize service-oriented data sharing, and differences in the middleware layer bring the need for integrating message and service-oriented architectures (SOAs) in order to unite information coming from both BIM and IoT nodes (Fig. 7.10).

The Solution In order to eliminate differences in communication approaches, protocols and middleware the data sharing approaches can integrated with wrapper Web services or at client level (these approaches are presented in the previous two patterns). In fact, there is another option where BIM entities can mimic the IoT nodes and present their states, or changes in their states using the lightweight protocols of IoT. The final pattern of the book presents an architecture where a series of virtual (soft) sensors are populated and used to represent information acquired from BIM, and publish this information to the message brokers where the BIM elements become the virtual 'Things' themselves. The pattern focuses on information acquisition and fusion. The architecture presented here is composed of five layers. The IoT nodes publish their states or change in their states to a message broker. On the other hand, virtual sensors that act as the observers of every building

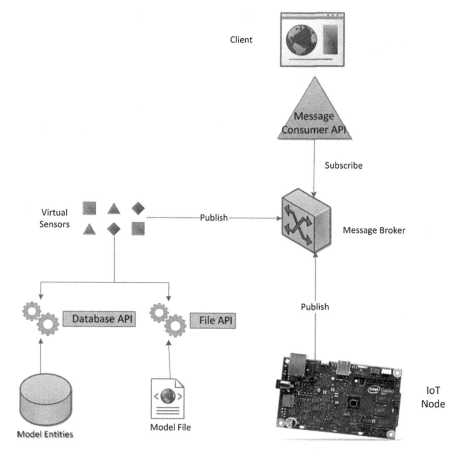

Fig. 7.10 Real-time BIM callback pattern. Grant of permission by Intel Corporation

element in the BIM periodically query the model entities, and once they notice a state change they publish a notification to the message broker. A message consumer API is the subscriber of the message broker, and once the changes are notified it will inform the client about the changes in the BIM and IoT nodes.

References

ASmarterPlanet: IBM Smarter Planet Video: Internet of Things. http://youtu.be/sfEbMV295Kk (2010). Accessed 10 Mar 2011

Campa, S.A., Rodrıguez-Gonzalez, A.B., Ramos, J., Caamano, A.J.: Distributed detection of events for evaluation of energy efficiency in buildings. In: Proceedings of IFIP International Conference on New Technologies, Mobility and Security (NTMS 2011). http://smartgrid.ieee.

org/publications-renewable-energy/4064-distributed-detection-of-events-for-evaluation-of-energy-efficiency-in-buildings. Accessed 17 Mar 2011

Guinard, D., Trifa, V., Mattern, F., Wilde, E.: From the internet of things to the Web of things: resource oriented architecture and best practices. In: Uckelmann, D., Harrison, M., Michahelles, F. (eds.) Architecting the Internet of Things, pp. 97–129. Springer, New York (2011). ISBN 978-3-642-19156-5

Kwon, J.W., Park, Y-M., Koo, S.-J., Kim, H.: Design of air pollution monitoring system using ZigBee networks for ubiquitous-city. In: Proceedings of International Conference on Convergence Information Technology (ICCIT 2007), http://www.computer.org/portal/web/csdl/doi/10.1109/ICCIT.2007.361. Accessed 20 Jan 2010

Tse, W.L., Chan, W.L.: A distributed sensor network for measurement of human thermal comfort feelings. Sens. Actuators: A **144**(2), 394–402 (2008)

Usländer, T., Jacques, P., Simonis, I., Watson, K.: Designing environmental software applications based upon an open sensor service architecture. Environ. Model Softw. **25**(9), 977–987 (2010)

Chapter 8
Summary and Future Outlook

Abstract There are significant advantages provided by the patterns presented, including provision of client-independent and lightweight software architectures by making use of loosely coupled nature of RESTful Web services and message brokers. The integration of Internet of things (IoT) and building information models (BIMs), consuming information acquired from both resources, provides various opportunities for different domains ranging from city management to construction management and emergency response. This chapter provides a summary of the book contents and underlines the opportunities provided by integration of information from BIM and IoT elements.

8.1 Overall Summary

The book provides architectural approaches for (i) utilizing building information models (BIMs) over the Internet and for (ii) enabling information fusion between BIMs and Internet of things (IoT) elements. A BIM can be defined as a digital representation of a building that contains semantic information about the building elements. The BIM keyword also defines an information management process based on the collaborative use of semantically rich 3D digital building models in all stages of a project's and building's lifecycle. A BIM is defined by its object model schema, Industry Foundation Classes (IFC) which is the most popular BIM standard (and schema) currently. Chapter 1 starts by providing definitions of BIM, provides the general characteristics of IFC models, elaborates on sharing/exchange of BIMs and on model views, and concludes by discussing the role of BIMs in enterprises. The first evolution of BIM was from being a shared warehouse of information to an information management strategy. Today, the concept of BIM is evolving from being an information management strategy to a construction management method. This change in interpretation of BIM is fast and noticeable.

© The Author(s) 2015
U. Isikdag, *Enhanced Building Information Models*,
SpringerBriefs in Computer Science, DOI 10.1007/978-3-319-21825-0_8

Transformation from BIM to BIM 2.0 focuses on enabling an (i) integrated environment of (ii) distributed information which is always (iii) up to date and open for (iv) derivation of new information. Chapter 2 starts with providing recent trends in building information modelling and later elaborates on technologies that will enable BIM 2.0. BIM-based management of the overall construction processes is becoming a major requirement of the construction industry, thus the final part of this chapter provides matrices that can be used as a tool for facilitating BIM-based project and process management. In domains where detailed semantic information coupled with detailed geometric representations is of key importance (such as city modelling, construction, aircraft industry, ship production and so on), information models that represent these domains (such as BIM) have a complex model structure. Chapter 3 provides generalized service-oriented design patterns to facilitate the management of information models of complex structure. The chapter starts by summarizing design principles of service orientation, and later provides service-oriented architecture (SOA) patterns for managing complex information models such as (but not limited to) BIM. The present can be regarded as the start of the IoT era. IoT covers the utilization of sensors and near-field communication hardware such as RFID or NFC, together with embedded computing devices. The devices can range from cell phones to RFID readers, GPS devices to tablets, embedded control systems in cars to weather stations. In an IoT environment, a door would have the ability to connect with the fire alarm, or a chair would communicate with home lights, or a car would communicate with the parking space. This book focuses on single-board computers (SBCs) as IoT hardware components for acquiring and presenting building and indoor information. Thus, Chap. 4 elaborates on different SBCs which can be used for this purpose. IoT architectures do not only consist of hardware. The hardware would need to have operating systems to work, and to implement communication protocols to communicate with different hardware and humans. Furthermore, the middleware components facilitate communication and exchange of information between these devices. Integration portals play an important role in combing and integrating information acquired from multiple devices and presenting this information to the users. In this regard, Chap. 5 provides detailed information on the software components of IoT platforms. Web services are the endpoints of the Web which enable interaction with web objects. Two styles of Web services exist today: Simple Object Access Protocol (SOAP) and REpresentational State Transfer (REST). The REST is often preferred over the more heavyweight SOAP because REST does not leverage much bandwidth. REST's decoupled architecture made REST a popular building style for cloud-based APIs, such as those provided by Amazon, Microsoft and Google. Chapter 6 starts with providing technical information about RESTful Web services. Following this, the chapter presents RESTful design patterns for facilitating BIM-based software and Web service architectures. The IoT approach proposes a global wireless sensor/actuator network composed of everyday devices such as home appliances, city furniture, mobile phones or vehicles. Everyday devices would either be publishers of information, or subscribers of information coming from other people and devices. The information provided by each device would only be limited by the device's sensing capabilities. In order to represent real-time information about buildings/indoors and in city models, the integration between information acquired from IoT devices and BIM elements (including door, window, wall, slab, roof, staircase and space) is an obvious necessity. Chapter 7 first presents patterns related to integration of information provided by IoT elements, and then focuses on patterns that are formulated for integration of BIMs with information provided by the IoT elements.

8.2 Future Outlook

There are significant advantages provided by the patterns presented such as establishing client-independent and lightweight software architectures by making use of loosely coupled nature of RESTful Web services and message brokers. The integration of IoT and BIMs, consuming information acquired from both resources, provides various opportunities for different domains ranging from city management to construction management and emergency response. Some of these opportunities are given below.

Client-Independent Architecture The information provided by provided system architectures can be presented, visualized and analysed by any client application that is able to consume Web services and messaging protocols. The client can either be a CAD/GIS that is used for managing and monitoring the energy infrastructure of a city or a monitoring portal that visualizes the levels of air pollution in a virtual globe. Additionally, the client can also be a mobile device or a tablet computer. The information gathered from the IoT nodes such as temperature, humidity, state of building elements (i.e. the window being open/closed), the occupancies in rooms, information regarding danger of fire and flood, EMF radiation levels in the building, the state of heating/ventilation/air conditioning (HVAC) systems, the state of elevators and escalators (i.e. in operation or not), oxygen and other gas levels will be available for the interested parties regardless of their hardware, operating system and the software they use. This loosely coupled nature of the architectures would bridge the key technologies for acquiring and presenting real-time building information.

Ubiquitous Monitoring The concept of ubiquitous monitoring has two dimensions. In the first dimension the building elements can be viewed as "things" that are continuously providing information about their own state (i.e. when their state changes). For example, a when a window becomes closed it will provide/publish information regarding its state of being closed, or when a room is on fire, the room will provide information about its state of being-on-fire; in summary, ubiquitous monitoring the of the role of information providing shifts from computers to "things" (i.e. in our case to the building elements or IoT nodes). In the second dimension, by the use of ubiquitous monitoring, the information regarding building elements would be available 24/7 regardless of the situation (i.e. emergency or non-emergency). City dashboard applications would be the main consumers of this ubiquitous information. By implementation of the described architectures, the information that is provided by the BIMs and IoT nodes will be available without interruption even when an emergency situation or disaster occurs, if the connectivity of the devices/hardware can be maintained. Although achieving this is not a simple task, the presented architectures provide unique opportunities for ubiquitous monitoring. For example, an earthquake can damage several floors of a building and thus IoT nodes in these floors might stop functioning; the situation would be similar to a flood situation. In this case, other nodes in the network will continue to function and provide information about the state of the elements and the condition of the building; i.e. as the functions of these are independent from others that are damaged and thus the nodes will continue to operate and report gas levels or a fire that would occur in the building following the earthquake. In this case, combining semantic information coming from building models with sensor information provides advantages in answering questions such as "Would you provide the average CO_2 level in the rooms which are not affected by the fire?", "Would you provide the number of doors which are open in the floors that are affected by the flood?" and so on, can be answered in real-time.

Crowd-Sourced Monitoring Crowd-sourced monitoring refers to monitoring of events and physical conditions by a vast number of people and devices that are in/near the region where the event occurs. This approach is extremely useful to enable ubiquitous monitoring of an emergency/disaster situation. When a disaster such as an earthquake occurs, a huge number of people will not be in a condition to monitor the event; in addition, a considerable number of devices and sensors would stop functioning. In this situation, if a crowd of people and IoT nodes controlled or monitored by them can provide information regarding the emergency situation, this information would be highly valuable. A recent example is the crowd-sourced monitoring of radiation levels in various regions of Japan after the Japan earthquake. Hundreds of real-time networked Geiger counter measurements (contributed by concerned citizens in Japan) are gathered and abstracted by Xively and visualized in a virtual globe. Crowd-sourced monitoring can be very useful in real-time monitoring of the spreading of a fire in a forest, i.e. IoT nodes located in the buildings in the forest area can provide information if the fire has reached the region by providing temperature and gas level information inside and outside the building. Crowd-sourced monitoring can also be useful in Tsunami events; a lot of houses can become flooded in a very short time period, and a crowd-of-nodes that are still in operation (in buildings that are not flooded) would provide key information regarding the invasion of flood water within the region.

Human–Building Interaction With the integration of IoT nodes and BIM entities, the building itself becomes the information provider. This information can either be provided in a continuous manner (i.e. according to publish/subscribe model) or on demand. The provision of information on demand by the building elements, when combined with the nodes' ability of controlling the actuators (i.e. devices for moving and controlling mechanism) offers opportunities by enabling human–building interaction. For example, in a fire response operation an emergency responder can acquire information from the sensors located on each floor regarding the spreading of the fire; in response, it can then invoke the Web services to interact with IoT nodes which will then invoke the actuators to close the doors on certain floors to prevent the spread of the fire to other floors. Furthermore, M2M autonomous interaction is also possible and a node can collect information regarding the emergency situation, and interact with another node to perform an action. This concept when implemented might be regarded as a shift from an automated building to an intelligent building. Human–building interaction might provide other opportunities in other emergency situations, such as floods; for example, sensors in the building can interact with the actuators to close doors to prevent some parts of the building from being flooded by water, in fact if there can be people in these parts of the building, they can be trapped as they cannot get out. In this situation, the people in the rooms can interact with the nodes (to control sensor and actuators) to let them out of that building part. In summary, the ability to consume information from sensors, and the ability to control the actuators provides unique opportunities by enabling human–building interaction in emergency response operations.

Printed in the United States
By Bookmasters